THE
WAR
FOR
KINDNESS

THE
WAR
FOR
KINDNESS

Building Empathy in
a Fractured World

Jamil Zaki

CROWN

NEW YORK

Published in the United States by Crown, an imprint of the Crown Publishing
Group, a division of Penguin Random House LLC, New York.
crownpublishing.com

CROWN and the Crown colophon are registered trademarks of
Penguin Random House LLC.

Library of Congress Cataloging-in-Publication Data
Names: Zaki, Jamil, 1980– author.
Title: The war for kindness : building empathy in a fractured world /
    Jamil Zaki.
Description: New York : Crown, 2019.
Identifiers: LCCN 2018046535 | ISBN 9780451499240 (hardback) |
    ISBN 9780451499257 (paperback) | ISBN 9780451499264 (ebook)
Subjects: LCSH: Empathy. | Kindness. | Social psychology. | BISAC:
    PSYCHOLOGY / Social Psychology. | PSYCHOLOGY / Emotions. |
    SCIENCE / Philosophy & Social Aspects.
Classification: LCC BF575.E55 Z35 2019 | DDC 152.4/1—dc23
    LC record available at https://lccn.loc.gov/2018046535

ISBN 978-0-451-49924-0
Ebook ISBN 978-0-451-49926-4

Printed in the United States of America

Jacket design and illustration by Michael Morris;
(fist graphic) CSA Images/Getty Images

10 9 8 7 6 5 4 3 2 1

First Edition

For Landon

CONTENTS

I WAS EIGHT years old when my parents began divorcing, but twelve by the time they finished. They were born ten thousand miles apart—my mother in southern Peru near the Chilean border, my father in the six-month-old nation of Pakistan. Twenty-five years later, Washington State University granted scholarships to students from the world's poorest nations, giving one to my mother. Around the same time, my dad's father gifted him a one-way ticket to the United States, and just enough money for a semester at WSU. They traveled from Lima and Lahore—each city about the size of Los Angeles—to the woodsy, sleepy town of Pullman.

Both of my parents felt disoriented in their new home. My dad had been middle class in Pakistan but was penniless by American standards. On many mornings, he'd buy three hot dogs for one dollar at a local restaurant, spreading them out over breakfast, lunch, and dinner. Breaking Muslim norms hurt, but he couldn't afford any other option. My mom was assigned a host family to ease the transition, but they lived eighty miles from campus. She spent much of her time alone, studying. WSU held a welcome reception for its international scholars. My dad showed up for the food, my mom for the company.

They married and moved to suburban Massachusetts, where I was born. But as they became more comfortable with the United

States, they grew less comfortable with each other. My father began a computer hardware company and worked eighteen hours a day. His American dream culminated in a beige Mercedes and a massive peach-colored stucco house, both of which struck my mother as grotesque. After not seeing much of my father for a few years, she decided to see even less of him.

As my parents receded from each other, they scorched the earth between them. Outside of court they studiously avoided contact. My dad would wait in my mom's driveway at a dedicated time each week, I would walk outside, and my mom would lock the door behind me, careful to not show herself. When I was thirteen, my father's mother died. That weekend when he arrived to pick me up, my mother walked outside and they hugged. It was the only time I remember them looking at each other in a ten-year span.

I shuttled back and forth between their houses but might as well have been moving between parallel universes—each defined by its own priorities, fears, and grievances. My mom is quintessentially Peruvian and values family above all else. She lost herself in anxiety over how the divorce would affect me, picking out signs that I was in pain and tallying those in a mental ledger of the damage my father had done. In my father's world, intellect and ambition mattered most. He often told me that where he came from, the student who scored highest on a big exam made it to college, but the kid who scored second best ended up on the street. When my grades slumped, he wondered aloud whether it would be worth the money to send me to college. He had broken his back to give my mother and me what he had never had, a favor we repaid by demoting him to half villain, half ATM. How could we not see that?

My parents each tried to conscript me into their war. They told me the secrets they were keeping from each other. They bought favor by letting me break the other's rules. They vented bitterly and, when I didn't join in, accused me of being on the other's side. I think all three of us believed that at some point I would have to choose one parent, giving up on ever really knowing the other.

In the classic 1983 film *WarGames*, a young Matthew Broderick hacks into "Joshua," an artificial intelligence program that, unbeknownst to him, is plugged into NORAD, the North American Aerospace Defense Command. He plays a simulation of thermonuclear war between the United States and the Soviet Union, nearly setting off World War III in the process. With Joshua set to take over NORAD's missile system and fire, Broderick convinces it to first try out every strategy. Joshua quickly realizes that no matter what either nation does, they both end up obliterated. "Strange game," the program reflects. "The only way to win is not to play."

So with my parents, I decided not to play—or at least not in the way they wanted. As they fought through me, I fought to hold on to both of them. Rather than picking a side, I tried to understand these two good people, who were trying to do right by me despite the pain they were in. While at my mom's house, I picked up the rules that governed her heart and mind, and made them true for myself. When I visited my dad, I adapted to his world. It was hard work. Like so many children of divorce, I was pulled in different directions by a centrifugal force. Sometimes it was hard to know what I believed. But I learned to tune myself to each of my parents' frequencies, and managed to stay connected to both of them, even as their ties to each other disintegrated.

When I think back on those days, I'm filled with gratitude. That two people's experiences could differ so drastically, yet both be true and deep, is maybe the most important lesson I've ever learned.

IMAGINE PUTTING ON a pair of goggles that work like thermal sensors but pick up emotion instead of body heat. You could watch, in glowing infrared, as anger, embarrassment, and joy bloomed inside people. If you kept watching, you would see that feelings do not stay put in any one person. When a friend cries in front of you or tells you a hilarious story, their voice and expressions leap through the air between you and into your brain, changing you in the process.

You take on their emotions, decode their thoughts, and worry about their welfare. In other words, you empathize.

Most people understand empathy as more or less a feeling in itself—*I feel your pain*—but it's more complicated than that. "Empathy" actually refers to several different ways we respond to each other. These include identifying what others feel (cognitive empathy), sharing their emotions (emotional empathy), and wishing to improve their experiences (empathic concern).[*]

I can't know for sure how you experience the color blue, let alone exactly how you feel when you're excited or frightened. Our private worlds circle each other in wobbly orbits but never touch. When two people become friends, their worlds inch closer together; when my parents split up, theirs drifted apart. Empathy is the mental superpower that overcomes this distance. Through it we voyage to others' worlds and make guesses about how it feels to be them. An impressive amount of the time, we get it right. Listening to a stranger tell an emotional story, we can describe how they feel with considerable accuracy. Glimpsing a face, we can intuit what a person enjoys and how much they can be trusted.

Empathy's most important role, though, is to inspire kindness: our tendency to help each other, even at a cost to ourselves.[†] Kindness can often feel like a luxury—the ultimate soft skill in a hard world. It puzzled Charles Darwin. According to his theory of natural selection, organisms should protect themselves above all else. Helping others did not fit into that equation, especially when we risked our own safety to do so. As Darwin wrote in *The Descent of Man*, "He who was ready to sacrifice his life . . . rather than betray his comrades, would often leave no offspring to inherit his noble nature."

---

[*] For a more detailed definition of empathy, see appendix A: "What Is Empathy?" on page 178.

[†] For further evaluation of major scientific claims in this book, see appendix B: "Evaluating the Evidence" on page 183.

In fact, kindness is one of the animal kingdom's most vital survival skills. Newborns are little bundles of need, and remain mostly helpless for days (geese), months (kangaroos), or decades (us). Either parents sacrifice to help them survive, or they risk leaving no offspring to inherit their selfish nature. The same goes for other kin: When an animal helps its relatives, it ensures the survival of its own genes. Unrelated animals can also benefit from acting kindly, especially when doing so builds alliances between them. Working together, they can find food, protect one another, and thrive in ways loners simply can't.

In these cases, kindness is smart, but that still doesn't explain why any one animal chooses to help another in a given moment. A mother squirrel doesn't know that her genes will be passed to the next generation, so why nurture her pups? A vervet monkey can't calculate the odds that a neighbor will return his favor, so why bother? Empathy is nature's answer to that question. When one creature shares another's emotions, seeing pain feels like being in pain, and helping feels like being helped.

Empathic experience undergirds kind action; it's a relationship far older than our species. A rat will freeze—a sign of anxiety—when its cage-mate is zapped with electric shocks. Thanks to that response, they also help each other, even giving up bits of chocolate to relieve the cage-mate's distress. Mice, elephants, monkeys, and ravens all exhibit both empathy and kind behavior.

In humans, empathy took an evolutionary quantum leap. That's a good thing for us, because physically, we're unremarkable. At the dawn of our species, we huddled together in groups of a few families. We had neither sharp teeth, nor wings, nor the strength of our ape cousins. And we had competition: Just thirty thousand years ago, at least five other large-brained human species shared the planet with us. But over millennia, we *sapiens* changed to make connecting easier. Our testosterone levels dropped, our faces softened, and we became less aggressive. We developed larger eye whites than other primates, so we could easily track one another's gaze, and intricate facial muscles that allowed us to better express emotion. Our brains developed to give us a more precise understanding of each other's thoughts and feelings.

As a result, we developed vast empathic abilities. We can travel into the minds of not just friends and neighbors but also enemies, strangers, and even imaginary people in films or novels. This helped us become the kindest species on Earth. Chimpanzees, for instance, work together and console each other during painful moments, but their goodwill is limited. They rarely give each other food, and though they may be kind to their troop, are vicious outside of it. By contrast, humans are world-champion collaborators, helping each other far more than any other species. This became our secret weapon. As individuals, we were not much to behold, but together, we were magnificent—unbeatable super-organisms who hunted woolly mammoths, built suspension bridges, and took over the planet.

As our species spread, so did our kindness. People share food and money with one another in cultures around the world. In 2017, Americans alone donated $410 billion to charity and spent almost eight billion hours volunteering. Much of this kindness flows directly from empathy. Highly empathic individuals donate more to charity and volunteer more often than their peers, and people who are momentarily inspired to empathize are more likely to help a stranger. And like a photonegative, our darkest times expose our noblest capacities, as with families who faced death to harbor Jews during the Holocaust, or teachers who shield their students during school shootings.

In his classic book *The Expanding Circle*, the philosopher Peter Singer claims that though we once cared for a narrow group of people—our kin, perhaps a few friends—over time, the diameter of our concern has expanded beyond tribe, town, and even nation. Now it encompasses the planet. The food we eat, the medicine we take, and the technology we use are sourced globally; our survival depends on countless people we will never meet. And we help people we will never know—through donations, votes, and the culture we create. We can learn intimate details about the lives of people half a world away and respond with compassion.

WE CAN, BUT we often don't, and this raises an important truth about empathy. Our instincts evolved in a world where most of our encounters were, in every sense, familiar. Friends and neighbors looked like us. Over a lifetime, we had countless chances to learn about their character—and they, ours. We had a future together, meaning that kindness and cruelty could be repaid. Karma was strong, direct, and unavoidable. Those we saw suffering were right in front of us, and by stepping in we could make a difference. These small, tightly knit communities were empathy's primordial soup, packed with ingredients that make caring easy.

Of course, we stepped in to help only certain people. The hormones that encourage parents to nurture children also made us suspicious of outsiders—potential rivals, cheaters, and enemies—and along with the ability to understand each other, humans developed a knack for separating ourselves into "us" and "them."

The modern world has made kindness harder. In 2007, humanity crossed a remarkable line: For the first time, more people lived in cities than outside of them. By 2050, two-thirds of our species will be urban. Yet we are increasingly isolated. In 1911, about 5 percent of British citizens lived alone; a century later that number was 31 percent. Solo living has risen most among young people—in the United States, ten times as many eighteen-to-thirty-four-year-olds live alone now than in 1950—and in urban centers. More than half of Paris's and Stockholm's residents live alone, and in parts of Manhattan and Los Angeles that number is north of 90 percent.

As cities grow and households shrink, we see more people than ever before, but know fewer of them. Rituals that bring us into regular contact—attending church, participating in team sports, even grocery shopping—have given way to solitary pursuits, often carried out over the Internet. At a corner store, two strangers might make small talk about basketball, school systems, or video games, getting to know all sorts of details about each other. Online, the first thing we encounter about a person is often the thing we'd like least about them, such as an ideology we despise. They are enemies before they have a chance to be people.

If you wanted to design a system to break empathy, you could scarcely do better than the society we've created. And in some ways, empathy *has* broken. Many scientists believe it's eroding over time. Think about how well each of these statements describes you, from 1 (not at all) to 5 (fits you perfectly).

I often have tender, concerned feelings for people less fortunate than I am.

I try to look at everyone's side of a disagreement before I make a decision.

For the past four decades, psychologists have measured empathy using questions like these, collecting data from tens of thousands of people. The news is not good. Empathy has dwindled steadily, especially in the twenty-first century. The average person in 2009 was less empathic than 75 percent of people in 1979.

When we *do* empathize, our care can be erratic. Consider the tragedy of three-year-old Alan Kurdi. In September 2015, Alan's family, having fled their native Syria, set out across a narrow strait in the Mediterranean Sea, hoping to make it from Turkey to Greece. Their rubber raft capsized among the waves, leaving them adrift in the dark for over three hours. Despite his father's desperate efforts, Alan, his brother, and his mother drowned. "I don't want anything," forty-year-old Abdullah Kurdi said the next day. "Even if you give me all the countries in the world, I don't want them. What was precious is gone."

After Alan's death, a photographer captured a devastating photo of his small body lying facedown on the shore. The image rocketed around the world, testifying to the humanitarian crisis. The *New York Times* reported, "Once again, it is not the sheer size of the catastrophe . . . but a single tragedy that has clarified the moment." Donations poured in to support Syrian refugees. Then, for the most part, people got on with their lives. The crisis raged on, but contributions and news coverage dropped just as quickly as they'd risen, all but disappearing by October.

Alan's death merited the wildfire of empathy it produced. So does the plight of countless other children in crisis. Yet we find it easier to empathize with single individuals—whose faces and cries haunt us—than the suffering of many. In laboratory studies, people express *more* empathy for one victim of a tragedy than they do for eight, ten, or hundreds.

It made sense for our ancestors to empathize with one person at a time, but that same instinct now fails us. We are inundated with depictions of suffering: hundreds of thousands of people died in Haiti's 2010 earthquake; as I write, eight million people in Yemen do not know where their next meal will come from. These numbers astound us but also leave us overwhelmed and eventually numb. Under their weight, our compassion collapses.

Tribalism creates even deeper problems; to see them in action, look no further than America's political wreckage. Fifty years ago, Republicans and Democrats disagreed on policy over dinner, but still ate together. Now each side sees the other as stupid, evil, and dangerous. Territories that were once neutral—from bathrooms to football fields—have turned into moral battlegrounds. Amid all this animus, people savor outsiders' pain. Trolls work tirelessly to provoke as much suffering on the other side as they can. In this bizarre ecosystem, care doesn't merely evaporate; it reverses.

It is no surprise, then, that empathy has become the focus of civic leaders, poets, and pastors—anyone trying to mend the social fabric. "There's a lot of talk in this country about the federal deficit," Senator Barack Obama said in a 2006 commencement speech at Northwestern University. "But I think we should talk more about our empathy deficit." Obama went on to lament that "we live in a culture that discourages empathy. A culture that too often tells us our principal goal in life is to be rich, thin, young, famous, safe, and entertained. A culture where those in power too often encourage these selfish impulses." According to him, recovering empathy is critical to healing the nation. The philosopher Jeremy Rifkin puts this in even starker terms, writing, "The most important question facing humanity is this: Can we reach global empathy in time to avoid the collapse of civilization and save the Earth?"

Since Obama and Rifkin expressed these concerns, things have gotten worse. Our culture is addled, stretched, and fraying at the seams. The same instincts that propelled kindness among insular groups planted the seeds for fear and hatred to grow as our world becomes larger and more diverse. News organizations and social media platforms profit from our divisions. Outrage is one of their products, and it is a growth industry.

Modern society is built on human connection, and our house is teetering. For the past dozen years, I've researched how empathy works and what it does for us. But being a psychologist studying empathy today is like being a climatologist studying the polar ice: Each year we discover more about how valuable it is, just as it recedes all around us.

DOES IT HAVE to be this way? That's the question I explore in this book.

For centuries, scientists and philosophers have argued that empathy is inherited through our genes and wired into our brains. I call this the Roddenberry hypothesis, because Gene Roddenberry enshrined it in the greatest television show of all time, *Star Trek: The Next Generation*. The USS *Enterprise*'s counselor, Deanna Troi, is known throughout the galaxy for her empathy. In contrast to her, Roddenberry gives us the android Data, who is blind to others' feelings (though excellent at violin and model-ship building).

The Roddenberry hypothesis contains two assumptions, each

part of a long intellectual tradition. The first is that empathy is a *trait*—something intrinsic to our personality that remains fixed over time. Deanna Troi is half human, half Betazoid—a telepathic humanoid race. Her empathy stems entirely from those alien genes: a gift of nature, with no nurture involved. A human could never aspire to be like her, any more than they could hope to breathe water or grow a tail. Data's lack of empathy was literally programmed into his positronic brain. The implication is that the rest of us, too, are coded by nature with a "level" of empathy—somewhere between Data and Troi. And like our adult height, we're stuck there for life.

This idea can be traced back to Francis Galton, a British scientist obsessed with measurement (his motto: "Whenever you can, count") and with human intelligence. In 1884, Galton joined forces with the London International Health Exhibition to open the world's first psychological testing fair. Londoners could make their way down a long, narrow table, taking a series of exams. At one station they responded as quickly as possible to flashes of light; at another, they tried to tell similar tones apart. Galton's tests failed to predict his subjects' intellectual ability or professional success, but he was unfazed, believing he simply needed better tools. Others agreed, and by the 1920s countless tests measured IQ, personality, and character.

Galton, Darwin's half cousin, was a fierce genetic determinist. He ranked ethnic groups by intelligence, invented the term "eugenics," and dreamed of a "utopia" in which people could be bred for intellect and moral worth. Eugenics, of course, aged poorly. But psychologists of his time were influenced by Galton's thinking, and many came to believe that their tests captured immutable "levels" of character. If you tested as moderately smart and very neurotic, that was how you had been born, and how you would remain until you died.

In the early twentieth century, psychologists began studying empathy. Their first instinct was to follow the testing trend, and they devised dozens of assessments. Some asked subjects to pick out emotions in faces. Others examined their responses to one another. How much did someone's heart rate jump when the person next to them received an electric shock? How sad did they become while listening to an orphaned child's story? Psychologists used these tests to sketch the typical "empathic person," who tended to be older, intelligent, female, and interested in art. Some hoped to use this information practically, for example to figure out who would excel as a therapist or judge. But the findings were less straightforward than they'd hoped. People who scored well on one empathy test didn't always score well on others. Some empathy tests predicted kindness; others did not.

Still, the testing trend continued, reaching its apex in 1990 with the concept of emotional intelligence (EI), created by the psychologists Peter Salovey and John Mayer. EI soon exploded into pop culture, and some pieces were lost in translation. Salovey and Mayer believed EI could be developed through practice, but gurus often claimed they could locate high-EI *people*, whom clients might want to hire or date. (Hint: If your partner doesn't like dogs, you might consider other options.) The implication: Someone's EI was a trait that could never change.

Deanna Troi might be preternaturally empathic, but for her, that's often a drag. In many *Star Trek* episodes, she runs into someone on the *Enterprise* and instantly crumbles—overtaken by their feelings. There's nothing she can do to turn her antenna off. By contrast, Data's hijinks often involve sarcasm, sadness, or romantic interest sailing past him. He plows into one faux pas after another, blissfully unaware. His lack of social grace is as involuntary as Troi's deep feeling.

This is the second part of the Roddenberry hypothesis: Not only is empathy an immutable trait, but our experience of it in any given moment is *a reflex,* instantaneous and automatic. This idea has its roots in an ancient assumption about how emotions work. In *Phaedrus,* Plato describes the human soul as a chariot. Its driver—symbolizing logic—struggles to control his horses. One of them represents emotion, and it does not come off well: a "crooked lumbering animal . . . with grey eyes and blood-red complexion . . . hardly yielding to whip and spur." Plato saw mental life as a war between reason and passion, one we often lose.

Not everyone agreed. The Stoic philosopher Epictetus believed that emotions were a product of thinking. "It is not events that disturb people," he wrote. "It is their judgments concerning them." This was an empowering view, because it meant that by changing how we think, we could change our feelings as well. Outside of the West, spiritual practitioners from Buddhist and other traditions perfected techniques to do just that.

Western thinkers, though, have favored Plato's perspective. They characterize feelings as ancient, animalistic impulses that appear unbidden, fueling bar fights, bad investments, and late-night ice cream binges. Early philosophers of empathy, such as Adam Smith, Theodor Lipps, and Edith Stein, claimed that empathy was likewise automatic: People can't help but take on one another's feelings (and if we're like Data, can't help but *not* take them on). This perspective grew into the modern notion that emotions are "contagious," spreading between individuals like a virus.

Around the time Salovey and Mayer first described EI, researchers in Parma, Italy, discovered empathy's biological roots entirely by accident. They were investigating how the brain controls movement, by placing pieces of food on a table in front of macaque monkeys. When the animal grabbed a treat, the researchers listened to its neurons fire through electrodes implanted in its skull. One day, they left their recording equipment on while an experimenter placed food on the table. As a monkey watched him, a burst of activity erupted inside its brain, even though the animal was sitting still. This was a confusing turn of events, but it happened again and again, in that monkey and then in others. The researchers dubbed these cells "mirror neurons," or, more informally, "monkey see, monkey do cells."

Hundreds of studies—some from my own lab—soon documented human mirroring: not just for movement, but for emotions as well. People who watched someone else feel pain, disgust, or pleasure activated the same parts of their brain they would while experiencing those states themselves. The punch line was simple and poetic. We really do feel each other's pain—and joy and fear. What's more, this physical manifestation of empathy seemed to inspire kindness. Those who mirrored others' pain also volunteered to be shocked in order to spare them; those who mirrored others' pleasure were more likely to share money with them.

The research wasn't always consistent. In some cases, mirroring failed to predict kindness or even how much empathy people felt. And it's not clear exactly how mirroring works in the brain (interested readers can turn to the notes for more on this). But some researchers were nonetheless convinced they had found the holy grail of human goodness. One neuroscientist, epitomizing the breathless mood of the time, referred to mirror cells as "Gandhi neurons." And to non-neuroscientists, fMRI scans made empathy feel solid. Brain images, with their mesmerizing colors, evoke the truth. People tend to believe claims about their own mind when those statements include even the thinnest reference to neuroscience.

Mirroring became the dominant account of empathy, and it fit perfectly with the Roddenberry hypothesis. Brain scans tempt people into thinking of their minds as "hardwired," rigidly built to work a certain way. This metaphor, inspired by computer science, suggests that we can no more change our mind than we can rearrange our organs.

FROM PLATO TO Galton to modern psychology and neuroscience to the pop culture miracle of *Star Trek*, the received wisdom is clear: Empathy is beyond our control. If it is a trait, then there's nothing we can do to become more empathic over time. And if it's a reflex, there's nothing we can do to change how much we feel for one another in the moment. This is all well and good when empathy comes naturally: for instance, among our family, friends, or tribe. But it's bad news for modern times. It means that whenever we fail to empathize, we've hit the limits of our circuitry. We must simply stand by and watch as our world becomes more callous and disconnected.

Thankfully, the Roddenberry hypothesis—and the centuries of thought it represents—is wrong. Through practice, we can grow our empathy and become kinder as a result.

This idea might sound surprising, but in fact it is supported by decades of research. Work from many labs, including my own, suggests that empathy is less like a fixed trait and more like a skill—something we can sharpen over time and adapt to the modern world.

Consider our diet and exercise habits. Humans evolved in an environment where exercise was constant and sustenance was scarce. In response, we developed a taste for fat, protein, and rest. Now many of us are inundated with fast food and rarely required to exert ourselves. If we allowed our instincts to take over, we could indulge ourselves into an early grave. But many of us don't accept this; we fight to stay healthy, adjusting our diets and going to the gym because we know it's the wise thing to do.

Likewise, even if we have evolved to care only in certain ways, we can transcend those limits. In any given moment, we can turn empathy up or down like the volume knob on a stereo: learning to listen to a difficult colleague, or staying strong for a suffering relative. Over time, we can fine-tune our emotional capacities, building compassion for distant strangers, outsiders, and even other species. We can free our empathy from its evolutionary bonds.

In other words, empathy is not a superpower after all, bestowed upon Betazoids and, to some extent, the rest of us. It's a regular old power, like being strong, agile, or good at Scrabble. Some people are genetically predisposed to be stronger than others—but strength is also up to us. Live a sedentary life, and your muscles will atrophy. Stay active, and they'll grow.

My parents' divorce was an empathy gym for me. It forced me to exercise compassion—to work at connecting with both my mother and father, instead of shutting down or engaging in their conflict. We can all choose to become more empathic, just like we can choose a healthier lifestyle. In many cases, they are the same choice. As the novelist George Saunders writes, "There's a confusion in each of us, a sickness, really: *selfishness*. But there's also a cure. So be a good and proactive and even somewhat desperate patient on your own behalf—seek out the most efficacious anti-selfishness medicines, energetically, for the rest of your life."

This book is about those medicines, and the science in which they are rooted. With the right treatments—including unlikely friendships, art, and community building—we can grow a more muscular kind of empathy, and broaden our kindness along the way. In these pages, we will meet cops who learn to interact more peacefully with civilians, Hutus and Tutsis moving toward forgiveness after genocide, and lifelong bigots dissolving their hatred. We'll see ex-convicts discussing novels with the judge who sentenced them, rediscovering their humanity in the process, and NICU physicians and nurses learning to help families through their hardest moments without drowning in their own pain.

Fighting for kindness is not easy for them, and it won't be easy for any of us. This book will not provide ten simple steps for how to be kinder today. It won't promise that, despite appearances, people are essentially good after all. Humanity might naturally be 39 percent kind, or 71 percent kind, or somewhere in between. What matters is not where we begin, but where we can go.

In five years, or one, the world could be a meaner place or a kinder one. Our social fabric could further tear or start to mend. We don't owe others empathy, especially if they meet us with cruelty or indifference. But if we succumb to our lazier emotional instincts, we will all suffer more. The direction we take—and our collective fate—depends, in a real way, on what each of us decides to feel.

CHAPTER 1

The Surprising Mobility of Human Nature

Eppur si muove (And yet it moves).

—ATTRIBUTED TO GALILEO GALILEI

A CENTURY AGO, almost everyone believed the ground lay still beneath us. Australia had always been an island, Brazil and Senegal had always been separated by the Atlantic; it was too obvious to discuss. Alfred Wegener changed that. Wegener was that not-so-classic combination of adventurer and meteorologist. He broke a world record by floating above Europe in a weather-tracking balloon for more than two days. He trekked across Greenland, detonating bombs in the tundra to gauge how deep the ice caps were. He would die on one of those trips at the age of fifty.

Studying maps of the ocean floor, Wegener noticed that the continents complemented one another like puzzle pieces. "Doesn't the east coast of South America fit exactly against the west coast of Africa, as if they had once been joined?" he wrote to a lady friend. "This is an idea I'll have to pursue." Wegener spotted other mysteries. The African plains were covered in scars left by ancient glaciers. If they had always been near the equator, how was that possible? Identical species of ferns and lizards were spread across Chile, India, and even Antarctica. How could they have traveled so far?

At that time, geologists believed that ancient land bridges once spanned the oceans, allowing life to crisscross between continents.

This did not satisfy Wegener. In his 1915 book, *The Origin of Continents and Oceans,* he proposed a radical alternative. The earth's land had once clumped together in a single mass—Wegener dubbed it "Pangea"—and for eons had rumbled apart into the continents we now know. The Atlantic Ocean was younger than people realized, and was growing. Animals that had evolved as neighbors had drifted to far-flung corners of the planet. The earth's surface was moving—imperceptibly, but constantly.

Wegener's idea did not land gently. Geologists ruthlessly mocked "continental drift," as it came to be called. Wegener was not part of their field, and insiders couldn't believe he had the gall to challenge their well-established notions, especially with such a strange idea. Summing up dozens of similar reactions, one researcher described continental drift as the "delirious ravings of people with bad cases of moving crust disease and wandering pole plague." A few came to Wegener's side, forming a small camp of geological "mobilists," but traditional "fixists" succeeded in defending a stationary earth. As Rollin Chamberlin, editor of the *Journal of Geology,* wrote, "If we are to believe Wegener's hypothesis we must forget everything which has been learned in the last 70 years and start all over again." At the time of his death, Wegener's theory had been tossed in the rubbish bin of scientific history.

Decades later, scientists discovered tectonic plates, masses larger than continents pushed along by currents of magma. The North American and Eurasian plates drift apart from each other about as fast as your fingernails grow; they've moved some three feet in my lifetime. Wegener, a scientific outsider with an unbelievable idea, had been right after all. Geology was rewritten to acknowledge that even things that appear still can move.

WE NOW ACCEPT that the earth and sky are forever changing, but our understanding of ourselves has proven more stubborn. Sure, we get old, our bones stiffen, and our hair turns gray, but our essence stays the same. Over the centuries, the supposed location of that essence has shifted. Theologians placed it in the eternal human soul; earthlier philosophers focused on natural character and virtue. In the modern era, human essence has become thoroughly biological, grounded in our genes and coded into our bodies.

No matter *where* human nature resides, it is assumed to be constant and immutable. I call this belief "psychological fixism," because it views people the way geologists once saw the continents. Fixism can be comforting. It means we can know who others truly are, and know ourselves as well. But it also limits us. Cheaters will always cheat, and liars will always lie.

Phrenology, a nineteenth-century "science," held that each mental faculty was housed in its own section of neural real estate. Phrenologists used calipers to measure the bumps and valleys on a person's skull, determining their degree of benevolence or conscientiousness. This sort of fixism was useful in defending prevailing social hierarchies. The phrenologist Charles Caldwell toured the American South arguing that people of African descent had brains built for subjugation. Others used supposed biological truths to argue that women were not worth educating, the poor were destitute for good reason, and criminals could never be reformed. As a science, phrenology was bankrupt, but as an ideology, it was convenient.

By the early twentieth century, neuroscience had outgrown phrenology, but there remained a lingering sense that our biology was fixed. Researchers knew that the human brain developed in leaps and bounds throughout childhood—not just growing, but reshaping into a dazzling, intricate architecture. Then, for the most part, it appeared to stop. Using the tools available to them, neuroscientists couldn't detect any changes after people reached adulthood. This jibed with popular notions about human nature, and became dogma. Scientists came to believe that cuts heal, but neurons lost to concussions, aging, or bachelorette parties were never replaced.

The father of modern neuroscience, Santiago Ramón y Cajal, described this idea: "In adult centers the nerve paths are something fixed, ended, immutable. Everything may die, nothing may be regenerated. It is for science of the future to change, if possible, this harsh decree."

But science did not need to change this decree, only to realize it was wrong. One of the first discoveries to lead the way came about thirty years ago, through the study of songbirds. Each spring, male finches and canaries learn new tunes to woo potential mates. Scientists discovered that as these birds built their repertoire, they also sprouted thousands of new brain cells a day. Over the years, researchers spotted new neurons in adult rats, shrews, and monkeys as well.

Skeptics still wondered whether adult *humans* could grow their brains. Then a breakthrough came from an unlikely source: the Cold War. In its early years countries tested their nuclear weapons regularly. Then, following the test ban treaty of 1963, they stopped. Levels of radiocarbon (^{14}C), an isotope produced by nuclear detonation, spiked and then plummeted just as quickly. Radiocarbon makes its way into the plants and animals we eat, and what we eat makes its way into new cells we produce. Neuroscientists such as Kirsty Spalding took advantage of this. Borrowing from archaeologists, Spalding "carbon-dated" brain cells based on their levels of ^{14}C, tagging the year they were born. Surprisingly, she found that people grow new neurons throughout their lives.

In other words, the brain is not hardwired at all. It changes, and these shifts are not random. MRI studies have now repeatedly shown that our experiences, choices, and habits mold our brains. When people learn to play stringed instruments or juggle, parts of their brain associated with controlling their hands grow. When they suffer chronic stress or depression, parts of their brain associated with memory and emotion atrophy.

Over the years, fixism has sprung other leaks. The more scientists looked for an unchanging "human nature," the less evidence they found. Consider intelligence. Francis Galton claimed it was baked into us at birth, never to budge. But a century later, in 1987, the psychologist James Flynn discovered a startling trend: Over the previous four decades, the average American's IQ had shot up by fourteen points. Other researchers have documented similar effects around the world in the years since. Crucially, intelligence changes even across generations of the *same* families. Such shifts are almost certainly not genetic in origin and instead reflect new choices and habits—for instance, in nutrition or education. Consistent with this idea, poor children who are adopted into more well-off households see their IQs rise by more than ten points. And in a recent analysis of over six hundred thousand people, psychologists found that for each year of schooling a person completed, their IQ increased by about a point, effects that last throughout their life.

Our personalities also change more than we might realize. After leaving home, new adults grow more neurotic. After getting married, they become more introverted; after starting their first job, they become more conscientious. We can, of course, also change intentionally. Psychotherapy leaves people less neurotic, more extroverted, and more conscientious than they were before—and these changes last at least a year after therapy ends. Personality doesn't lock us into a particular life path; it also reflects the choices we make.

THE SCIENCE OF human nature can now take a page from geology and finally reject fixism. We're not static or frozen; our brains and minds shift throughout our lives. That change might be slow and imperceptible. And yet we move.

In a nod to Wegener, we might call this idea "psychological mobility." Mobility doesn't mean anyone can be anything. Try as I might, I'll never move objects with my mind or win a Nobel Prize in physics. Our genes absolutely play a role in determining how smart, neurotic, and kind we are, and there's no denying that they're fixed at birth. Human nature is part inheritance and part experience; what's up for debate is how much each part matters.

Consider intelligence. A person's genes might predispose him to be relatively high or low in smarts. We can call this their "set point." But each person also has a range. Their intelligence registers as higher or lower depending on who raises them, how long they go to school, and even the generation into which they are born. A fixist focuses on a person's set point, asking how smart a person *is*. A mobilist focuses on ranges, and asks how smart that person *can be*. Both of these questions are important, but fixism has dominated more conversations about human nature than it deserves. As a result, we've underestimated our power over who we become.

According to the Roddenberry hypothesis, empathy is a trait, locked away and impervious to our efforts. This idea jibes with common sense. Of course some people care more than others; that's why we have saints and psychopaths. But what do these differences mean?

Imagine two people: Saul and Paul. Paul is less inclined toward empathy, more toward selfishness. A fixist would argue that this will hold true forever, because people seldom depart from their set point. Here, that idea is depicted by both men having a relatively narrow range. Even on his best day, poor Paul can barely empathize as much as Saul on his worst:

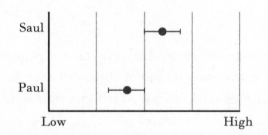

This perspective has some truth to it. Empathy is at least somewhat genetic, as demonstrated by studies of twins. In some of this research, twins decipher people's emotions based on pictures of their eyes; in others, their parents report on how often each twin shares her toys with other kids. In one particularly creative experiment, researchers visited two-to-three-year-old twins in their homes. One scientist pretended to accidentally slam her own hand in a briefcase. A second secretly measured how concerned children became and how much they tried to help their injured visitor.

No matter what the measure, identical twins tend to be more similar than fraternal ones. Both types of twins share a household, but identical twins share all their genes instead of half. The extent to which they "look" more alike than fraternal ones—in their personality, intelligence, and so forth—is what scientists chalk up to heredity. Analyses like this suggest that empathy is about 30 percent genetically determined, and generosity closer to 60 percent. These effects are substantial—comparable to IQ's genetic component of around 60 percent. And they are stable. In one study, people completed empathy tests several times over twelve years. If you knew a person's score at twenty-five, you'd do a decent job predicting how they'd perform at thirty-five.

While granting the importance of set points, a mobilist would argue that people can still change in meaningful ways. Take another look at the twin research we just discussed. Yes, empathy and kindness are partially genetic, but there is still room for nongenetic factors—experiences, environment, habits—to play a crucial role. Here, that elasticity is depicted by increasing the range of Saul and Paul's possible empathy. Depending on their experiences and choices, each one can travel quite a distance along his range. In this formulation, even if Paul's set point is lower than Saul's, his most empathic moments will surpass Saul's least:

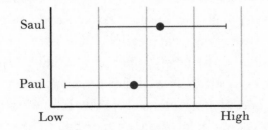

We now have decades of evidence demonstrating that empathy is shaped by experience. One-year-old children whose parents express high levels of empathy show greater concern for strangers as two-year-olds, are more able to tune into other people's emotions as four-year-olds, and act more generously as six-year-olds when compared to other children their age.

Nurture matters even more for children at the greatest risk. Psychologists have examined orphans in Romania, a country infamous for its maltreatment of institutionalized children. Kids here were often underfed and neglected. Never having been cared for, many Romanian orphans never learn to care about others, and display empathic deficits similar to those found in psychopaths. Some orphans, though, are lucky enough to be adopted by foster families, typically at two years of age. These children are spared many of their peers' problems, developing typical levels of empathy—especially if their foster parents respond warmly to them. A cruel environment moved these children to the left of their range, but swapping in a kinder one brought them back.

Circumstances mold empathy well into adulthood. A bout of depression, for instance, predicts that a person will become less empathic over the following years. More acute suffering also shifts empathy, in surprising and varied ways. When people cause it, their empathy erodes; when people endure it, their empathy grows.

We can't always avoid inflicting pain on others. Oncologists are constantly delivering unwelcome news: a patient's cancer has worsened; his treatment has failed; this illness will end her life. In 2017, U.S. managers laid off about thirty-four thousand employees a month. The psychologists Joshua Margolis and Andrew Molinsky call these moments "necessary evils." It's easy to sympathize with cancer patients and the newly unemployed, but people who carry out necessary evils suffer as well. For instance, about 50 percent of oncologists report feeling intense heartbreak and stress every time they break bad news. In laboratory experiments, even *pretending* to do so drove up medical students' heart rates.

When someone suffers at your own hands, caring for them can quickly give way to despising yourself. The resulting guilt takes a toll. During periods of heavy layoffs, managers who swing the ax develop sleep and health problems. In situations like these, people protect themselves by removing emotion from the equation. Margolis and Molinsky found that about half of individuals who performed necessary evils pulled away from the people they harmed. During layoffs, managers avoided thinking about their employees' families. They used curt language and cut off conversations. Doctors who had to deliver bad news focused on the technical side of treatment, trying to elide patients' pain.

To live with themselves, people who harm others often blame or dehumanize their victims, a process known as "moral disengagement." In the 1960s, one group of psychologists asked subjects to shock another person repeatedly. The subjects reacted by denying that the shocks hurt, and even thinking of victims as less likable. In a more recent study, white Americans were asked to read about the massacre of Native Americans at the hands of European settlers. Afterward, they doubted that Native Americans could feel complex emotions such as hope and shame.

Disengagement builds emotional calluses. For decades, the psychologist Ervin Staub has studied individuals who kill during war or genocide. He finds that they turn off their empathy, "reducing [their] concern for the welfare of those [they] harm or allow to suffer." In 2005, researchers interviewed death workers at prisons in the American South. Consistent with Staub's view, executioners claimed that death row inmates had "forfeited the right to be considered full human beings." Workers who were most involved in killing—for instance strapping inmates to gurneys for injections—dehumanized them the most. The closer executioners came to their victims, the less they saw them as people.

Causing pain can move people to the left of their empathic range, making caring harder, but people who endure great suffering often become *more* empathic as a result. Traumas—including assault, illness, war, and natural disaster—are psychological earthquakes that rock the foundations of people's lives. Survivors see the world as more dangerous, crueler, and less predictable after trauma than they did before. Many suffer from post-traumatic stress disorder: overtaken by flashbacks of their worst moments, fighting uphill to get their lives back. But most people who endure trauma do *not* develop PTSD. Six months after the fact, less than half of sexual assault victims report symptoms; among combat veterans, that number is about an eighth.

Trauma survivors who are supported by others typically have an easier path to recovery. Afterward, they often turn around and play that role for others. After Hurricane Harvey struck Houston, a group of Katrina survivors known as the "Cajun Navy" hauled dozens of boats to Texas to help recover victims. Thousands of other trauma survivors have switched careers to become "peer counselors"—helping others heal from wounds they once suffered. Veterans talk each other through hopeless moments. Addicts who have been clean for ten years help others make it through the first ten days.

Psychologists call such kindness "altruism born of suffering," and it is everywhere. Researchers recently examined war-torn communities in more than forty countries, including Burundi, Sudan, and Georgia. People in these towns, villages, and cities underwent unthinkable pain, and could be forgiven for retreating from public life. Instead, they redoubled their commitment to social movements and civic engagement. When researchers gifted them with money, they were more likely to share it with each other than people in unaffected towns. Likewise, victims of political violence and natural disasters volunteer at unusually high rates to help homeless people, the elderly, and at-risk children. And 80 percent of rape survivors report deeper empathy in the months after being attacked than they did before.

These positive effects last years. In one study, the psychologists Daniel Lim and Dave DeSteno measured the number of painful life events individuals had experienced throughout their lives—things like car accidents, severe illness, or victimization by crime. These people then came to the lab, where they met another participant who was struggling with a frustrating task. Participants stepped in to help this other person, and those who had suffered most were the most helpful, even if their painful experiences had occurred long before.

When survivors help others, they also help themselves. "Victims" are often stereotyped as weakened by trauma, but many emerge stronger and more fulfilled. "Post-traumatic growth"—including greater spirituality, stronger relationships, and a renewed sense of purpose—is almost as common as PTSD. Survivors who feel deepened empathy and act on it are most likely to report post-traumatic growth. When they counsel new survivors, they realize how far they've come, and how strong they are. And if the pain they endured can help them help others, it was not for nothing. As the great psychologist and Holocaust survivor Victor Frankl writes, "A man who becomes conscious of the responsibility he bears toward a human being . . . will never be able to throw his life away. He knows the 'why' for his existence, and will be able to bear almost any 'how.'"

EXPERIENCES CAN AND do move us along our empathic range, but the changes we've seen so far happen by accident. People don't harm others *in order* to care less; they merely adapt to the choices they've made. Victims certainly don't choose to be harmed; they grow kinder as a result of their circumstances. Few people choose their families, and no one chooses their genes. A second question, then, is not whether empathy can grow or shrink, but whether we can change it on purpose.

One heartening piece of evidence is that simply *believing* it is possible to change one's empathy helps to make it so. I learned this from one of my intellectual heroes, Carol Dweck.

I first met Carol during my job interview at Stanford. I'm a nervous person to begin with, and that was a nervous day for me, so talking with her was enough to tip me into a low-grade panic. On my way to her office, I stopped in the bathroom, soaked paper towels in cold water, and held them to the back of my neck, hoping to slow my inevitable sweating. During our meeting I talked at a mile a minute. My big question, I said, was whether people can change how empathic they are. I told her what you've just read: that empathy is partially baked into our genes but also changes with circumstances.

"But what do people think?" she asked.

I was confused. I had just rattled off a tight five minutes on what scientists think. Perhaps I had been unclear? Rushed? Un-hirable? I began rehashing my summary, but Carol interrupted me.

"No, I mean what do *people* think? Not researchers, but the people in their studies."

Researchers are, of course, people, too. But Carol's deceptively simple question made me realize that I had rarely considered what non-scientists believe about empathy.

This matters, and if anyone would know why, it's Carol. Over several decades, she has studied "mindsets," or what people believe about their own psychology. Carol finds that people fall into two general camps. "Everyday fixists" believe that pieces of our psychology, such as intelligence and extroversion, are unchangeable traits. Fixists define people, including themselves, by their set point. "Everyday mobilists" think of these same qualities more like skills. Sure, they might have a certain level of intelligence now, but that can shift, especially if they work at it.

Mindsets affect what people do, especially when things get tough. In a famous set of studies, Carol and her colleagues first measured students' mindsets about intelligence. Students then completed a challenging problem set, and later learned that they'd performed poorly. Fixist students attributed their failure to a lack of ability. And when they failed academically, they shied away from opportunities for extra training. To them it might have seemed pointless—if I can't improve, why try?—and embarrassing to boot. By accepting remedial education, they were admitting they were not smart and never would be. Mobilist students, on the other hand, embraced opportunities for additional learning. They reasoned that the more work they put in, the more they'd grow.

Carol not only measures mindsets; she also changes them. She and her colleagues have students read essays suggesting that intelligence is malleable. No matter how they start out, these students *become* mobilists, and try harder at intellectual tasks as a result. This kind of change can produce long-term effects. In a recent review of more than thirty studies, psychologists found that people who are taught that they can become smarter end up with slightly—but reliably—higher GPAs in the following school year. Mindsets are especially powerful in bolstering the performance of minority students and, in some cases, decreasing racial achievement gaps.

Once I arrived at Stanford, Carol and I—along with our colleague Karina Schumann—decided to see whether empathy might work in a similar way. We reasoned that people who believed it was a trait would avoid it during tough moments. People who believed it was a skill, on the other hand, might dig in, trying to empathize even when it was hard.

We began simply, by asking hundreds of people to pick which one of these statements they agreed with:

In general, someone can change how empathic a person they are.

In general, someone cannot change how empathic a person they are.

Our participants were split down the middle, about half fixists, half mobilists. With this information in hand, we ran them all through an empathic obstacle course: a series of circumstances in which empathy often fades. In most cases, mobilists tried harder than fixists.

For instance, they spent more time listening to emotional stories told by someone of a different race, and said they would devote more energy to considering the opinions of someone on the other end of the political spectrum.

Carol, Karina, and I also *changed* people's view of empathy, by presenting them with one of two magazine articles. Both started with the same passage:

> Recently, I bumped into someone I went to high school with over 10 years ago. As with all post–high school encounters, I couldn't help but compare the person in front of me to the person I remembered. Mary was one of those unsympathetic types who didn't really ever put herself in other people's shoes or understand how other people felt.

The fixist article continues:

> Can you imagine my lack of surprise to find that she is now a mortgage lender who sometimes repossesses the homes of struggling homeowners? Meeting such a similar person now, I wondered why Mary hadn't changed—why hadn't she grown out of her non-empathic persona?

It goes on to describe empathy as a trait, and closes with this reflection on Mary: "I guess it's no surprise that her level of empathy hadn't changed over time. Even if she had tried to learn to feel empathy for others, she probably would have been unsuccessful because it is just a part of who she is."

The mobilist article takes a different turn:

> Can you imagine my surprise to find that she is now a social worker with a family and an active role in community service? Meeting such a different person now, I wondered how Mary had changed so much.

This article then describes empathy as a skill, providing evidence that people *can* and do grow their capacity to care. It wraps up, "I guess she worked at developing feelings of empathy over the years. Now, as a social worker, she can pass on the message to others: People can change how much empathy they feel for others."

People in our study believed each article. After reading that empathy was a trait, they agreed with fixist statements. After reading it was a skill, they became mobilists. Crucially, these beliefs changed their own choices. "New fixists," empathized lazily, for instance with people who looked or thought like them, but not with outsiders. New mobilists, by contrast, empathized even with people who were different from them racially or politically.

Mobilists flexed their empathy in other tough situations, too. In one study, Stanford students learned about a cancer awareness campaign on campus. We told them they could help out in a number of ways. Some of those ways promised to be breezy, for instance participating in a walkathon to raise research money. Others would be painful, for instance sitting in on a cancer support group and hearing sufferers' stories. We asked students how many hours they'd be willing to volunteer for each. New fixists and new mobilists pledged equal amounts of time when helping was easy, but mobilists volunteered more than twice as many hours than fixists when more was required. Situations that normally turn people away no longer fazed them.

Participants were randomly selected to read either the fixist or mobilist article, meaning that those in each group almost certainly didn't differ much from one another when they arrived at the lab. But in just a few minutes, we pushed them to the left or right of their empathic range.

This work highlights a deep irony. The Roddenberry hypothesis dominates our culture's view of how empathy works. In essence, we've all been living like fixists. There are enough barriers to modern kindness, and by imposing fixism on ourselves we've added another one. If we can break this pattern and acknowledge that human nature—our intelligence, our personality, and our empathy—is to some extent up to us, we can start to live like mobilists, opening up new empathic possibilities. Perhaps in reading this far, you've moved yourself in that direction already.

But can we do more than simply change our mindset? Can we take precise control over our experience in the moment, empathizing in just the way we want, when we want to? And if so, how?

CHAPTER 2

Choosing Empathy

RON HAVIV AND Ed Kashi witness pain for a living. "We gravitate toward the things that most people would run away from," Kashi reflects. For decades, these two photojournalists have documented funerals, uprisings, and everything in between. They each provide an unblinking look at their subjects' hardest moments. But the way they approach their work is different. Consider these two photos, taken about six thousand miles and two years apart:

Haviv captured the one on the left. During the Kosovo War, Serbs carried out an ethnic cleansing campaign, wiping out or expelling Muslims. Many fled to the nearby mountains, but shelter was sparse. "This community was living in the mountains and it was getting cold," Haviv explains, "and this child died of exposure.

Here they're preparing the child for burial, and in the Islamic tradition you wash the body before burying it."

To capture images like this, Haviv removes his own feelings from the equation. "I have a responsibility to be there for the public, not myself . . . not to become so emotional that I am unable to photograph." Surrounded by grief, he keeps a poker face.

Kashi took the photo on the right. The woman at its center, Maxine, had reached the last stage of Parkinson's and Alzheimer's. "It was so clear this was the day she needed to die," Kashi recalls. "Her husband had to tell her to let go, and it became clear that I was the one who should tell him to go do that. So, I got on my knees and said, 'Art, I think you need to tell Maxine that it's okay to go.' . . . He got up and did it, and an hour and a half later she was dead."

Unlike Haviv, Kashi dives headfirst into his emotions. "Some of the greats would say, 'I'm there to do my work,' and they're more dispassionate," he acknowledges, "but often I'm crying a lot in these situations." He cultivates intimate connections with his subjects. In Maxine's last months, she and Art had slept in adjacent beds, hers provided by the hospital. The night she died, they replaced her bed with a cot; Kashi slept in it so that Art wouldn't be alone.

Why do Kashi and Haviv work so differently? One possibility is that they can't help it. This is the second part of the Roddenberry hypothesis: Empathy is a reflex that washes over us when we encounter someone else's emotions. Kashi might be the Deanna Troi of photographers; Haviv might be more like Data.

Neither photographer would agree. Haviv cares enormously for his subjects. He also feels that to help them, he must stay collected enough to document their suffering. Later, when the job is done, he lets himself go. "I've trained myself so that I can become emotional once I'm away from the situation," he says. "Back at the hotel, I can cry." For Kashi, emotional connection is part of the job. "I'm almost in the role of a social worker."

Haviv and Kashi don't see themselves as controlled by their feelings. Each, in his own way, empathizes with purpose.

IMAGINE A TUG-OF-WAR. Two teams, red and blue, pull at a rope from either side. You could describe this event in many ways: a row of young faces and bodies contorted with effort, an ancient contest of wills, a defunct Olympic event petitioning for a comeback in 2024. A physicist could describe it differently: Each player exerts a force—a combination of their strength and the direction in which they pull. Each team's force can be drawn as a series of arrows, with the red team's pointed east (let's say), and the blue team's pointed west. Longer arrows correspond to players who exert more force, shorter arrows to those who exert less. If the total eastward force is greater than the westward force, the red team will inch toward victory. If the red team tires and the blue team rallies, these forces will shift and the tide will turn.

The psychologist Kurt Lewin saw human behavior in the same way. In the 1930s, Lewin built his grand theory on the principles of physics. He argued that people are governed by a set of psychological forces, or motives. We are pulled toward certain actions by *approach motives*, and pulled away from them by *avoidance motives*. If approach motives outweigh avoidance motives, we act; if not, we don't. Here he depicts—of all things—the process of buying groceries.

BUYING CHANNEL

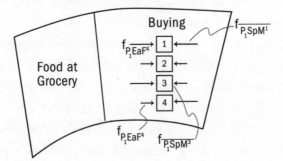

The shopper encounters four foods. Lewin depicts the motives enticing him toward each—maybe it's delicious or healthy—through a rightward arrow, and the motives pulling him away from it—maybe it's expensive or gluten-free—through a leftward one. Food 3 has many qualities that attract the shopper, and few that repel him. *Sold.* Food 4 is also a no-brainer, but in the other direction. Foods 1 and 2 are trickier. Food 1 attracts and repels at the same time; maybe it's delicious but also expensive (think filet mignon). Food 2 is not much of either, for instance, cheap but bland (maybe bologna). Both make for difficult choices: Food 1 because of conflict, Food 2 because of apathy.

Lewin used this theory to explain peer pressure, political turmoil, and everything in between. According to him, each choice reflects a tug-of-war in your mind. Everything you do, from getting out of bed to reprimanding your child to going for a jog, happens because the forces pulling you toward that action defeated the forces pulling you away from it.

What about how you feel? Until recently, most scientists didn't think of emotions as the result of a Lewin-style push and pull, or as choices at all. In 1908, the psychologist William McDougall argued that feelings were "instincts": ancient and preprogrammed. According to him, you don't decide when you feel fear, lust, or rage, any more than you choose whether your knee jerks when it's tapped with a mallet. Many people today agree with him. Researchers recently asked more than seven hundred people how they thought emotions work. About a third of them agreed with the statement "People are ruled by their emotions." About half believed that "emotions make people lose control."

McDougall also saw empathy as an instinct, triggered automatically by other people's feelings. "Sympathetic pain or pleasure," he wrote, "is immediately evoked in us by the spectacle of pain or of pleasure. . . . We then act on it because it is our own pain or pleasure." This view lives on in the Roddenberry hypothesis.

McDougall believed our empathic instinct was a positive force, the "cement that binds animal societies together." But a darker view has prevailed for centuries. In 1785, Immanuel Kant wrote that "good natured passion is nevertheless weak and always blind." In other words, even our most positive reflexes are still reflexes, and we can't determine when they're triggered. Empathy fires in response to a friend's pain, but not a stranger's. It fires in response to people who look like us, but not outsiders; to pictures, not statistics.

According to some, this is empathy's fatal flaw. Paul Bloom, a psychologist and the author of *Against Empathy: The Case for Rational Compassion*, writes, "Empathy's narrow focus, specificity, and innumeracy mean that it's *always* going to be influenced by what captures our attention, by racial preferences, and so on" (italics mine). When it misfires, we are helpless to redirect it: It is doomed to be biased, shortsighted, and ill-suited to the modern world. Bloom believes that to become truly moral beings, people must give up on feeling altogether, replacing it with a more Data-esque, rational benevolence. As he writes elsewhere, "Empathy will have to yield to reason if humanity is to have a future."

But, of course, feeling and reason are in constant dialogue. Emotions are *built* on thought. Someone who sees a bear might feel curiosity or terror, depending on whether he's at the zoo or in the woods. A child who falls down looks to her parents. If they respond calmly, she bounces back up; if they panic, she dissolves into a puddle of tears. Emotions reflect not just what happens to us, but how we *interpret* those things. The Stoic Epictetus knew this, and so did Shakespeare. As Hamlet puts it, "There is nothing either good or bad, but thinking makes it so."

This has a powerful consequence: By thinking differently, we can choose to feel differently. My colleague James Gross has examined this phenomenon for more than two decades. In dozens of studies, he's shown people emotional images like the ones at the top of this chapter. He asks subjects to turn down their feelings (like Haviv) or turn them up (like Kashi) by rethinking what they see. While looking at Maxine on her deathbed, someone might ramp up their sadness by thinking of Art, drinking coffee the next morning, without her for the first time in fifty years. Someone who wants to feel less sad might instead focus on the love they shared. After detaching, people in Gross's studies report weaker emotions. Their bodies show fewer signs of stress, and parts of their brain associated with emotional experience calm down. After they ramp up their feelings, the opposite happens.

This is one form of what I call "psychological tuning"—the quick, agile ways we alter our experience, for instance by focusing intently on a math problem, or rethinking what we feel. Tuning helps us constantly, especially in heated situations. Spouses who rethink their emotions during spats are more satisfied in their marriage. Israelis who do the same while reading about Palestinians' actions in Gaza advocate for more peace-positive policies.

People can choose not just to turn their emotions up or down but also to cultivate particular feelings that are useful in the moment. Happiness won't help a boxer going into the ring, but anger might. For a beggar seeking others' sympathy, sadness is wiser than fear. The psychologist Maya Tamir has found that people gravitate toward handy emotions, even when they feel bad. In her studies, people choose angry music to psych themselves up before a hostile negotiation, but sad music before asking for a favor. Emotions really *do* work like Lewin's tug-of-war: Whether you realize it or not, you're constantly weighing the costs and benefits of sadness, or joy, or anxiety, and choosing the feelings that serve your purpose.

Empathy is no different. Yes, it *can* occur automatically, like McDougall suggested. But more often, we choose or avoid it depending on whether it seems useful. There are obvious reasons to choose it. For one, it can feel good, because positive emotion is contagious. We are lifted up by the happiness around us like race cars drafting off one another's momentum. Empathy also feeds our deep-seated need for relationships. During my childhood, it was a way to feel closer to my parents amid family turmoil, so I worked at it. Likewise, when people want to connect—for instance, when they're around attractive or powerful individuals—they turn up their empathy and read others more clearly as a result.

Even when empathy doesn't feel good, we know it can make us *look* good. If Mother Teresa, the Dalai Lama, or Jesus are any indication, compassion and generosity are the clearest signs of virtue. When people must establish their moral bona fides, they turn to empathic actions. Individuals are more generous in public than in private, and also act kindly to convince *themselves* of their own goodness. In several studies, psychologists have put people under "moral threat," for instance, asking them to remember times they betrayed someone else's trust. To compensate, these participants help strangers, donate to charity, and advocate for environmentally friendly behaviors more than people who were not threatened.

But for every reason to choose empathy, there is another reason to avoid it. When others are in pain, connecting with them risks our own well-being. One friend of mine, who is a therapist, does her best not to schedule depressed patients for the end of the day, to avoid bringing their darkness home with her. In the 1970s, the psychologist Mark Pancer tested whether people would literally steer clear of painful empathy. He positioned a table in a busy student union at the University of Saskatchewan, posted information about a charity on it, and tinkered with its appearance. Sometimes the table was unmanned; other times a student sat by it in a wheelchair. Sometimes it contained an image of a smiling, healthy child; other times it showed a sick, sad one. The wheelchair and the sad image were triggers for empathy. Pancer found that students walked in a wider arc when it contained empathy triggers than when it didn't, keeping difficult feelings at bay.

When our own time or money is at stake, empathy feels like even more of a burden, and we avoid it more forcefully. A New Yorker walking the streets of Manhattan is inundated with suffering and need. If he takes it all in, he ends up in a double bind. He can give to others until he has nothing left, or live with the guilt of not giving. People often avoid empathy in situations like this. In one study, people who believed they'd later have a chance to donate to a homeless person avoided hearing a version of his story that included emotional details. It's not that they couldn't feel for him; they actively *chose* not to.

Even otherwise caring people become callous when they feel overwhelmed. The psychologists John Darley and Dan Batson once asked Princeton seminary students to prepare a sermon on the biblical parable of the Good Samaritan. It tells the story of a man traveling from Jerusalem to Jericho who is robbed, beaten, and left for dead by criminals. Luckily, a resident of Samaria later stumbles upon him. As described in the book of Luke, the Samaritan "had compassion, and went to him and bound his wounds, pouring on oil and wine . . . and took care of him." You might not want to actually treat wounds with wine, but seminarians still got the story's point and wrote about the power of caring.

Darley and Batson then instructed them to walk to another building to deliver their speech, but they added a twist. They told some students that their sermon would not begin for a while, and that they could take their time. Others learned time was tight. Students ambled (or sprinted) across Princeton's manicured grounds, and as they reached the building, encountered a man slumped in a doorway. As the students neared him, he coughed and groaned. In fact, he was an actor, secretly recording how they responded. Over 60 percent of them helped when they were in no hurry, but only 10 percent did when they felt rushed. The irony here is palpable: Students wouldn't help a man lying on a sidewalk because they were in too much of a hurry to give a speech about how important it is to help a man lying on a sidewalk.

People who avoid empathy often hurt themselves in the process. Decades of evidence demonstrate that individuals who empathize with others also help themselves: attracting friends more easily, experiencing greater happiness, and suffering less depression than their less empathic peers. When someone decides they don't have the resources or energy for other people, they deprive themselves of those benefits. In one study, the psychologist John Cacioppo and his colleagues surveyed people annually for ten years. Individuals who were lonely in a given year reported being more self-centered the following year. Self-centeredness, in turn, predicted deeper loneliness and depression in the future. Lonely individuals' motives were off base—empathy felt like it would overwhelm them, so they focused on themselves and ended up worse off.

BORROWING FROM LEWIN, we can see the problems of modern kindness in a new light. When empathy evolved, humans were enmeshed in close relationships. We had reason to care about almost everyone we saw. These forces pulled us toward empathy and made it easy. Now we are isolated, stressed, and drowning in animosity. We have more reasons to avoid empathy than ever.

Restating a problem is not the same as solving it. But in this case, it offers some ideas. Lewin famously wrote, "There is nothing so practical as a good theory." Describing water currents is an academic exercise; diverting them to irrigate crops is a technological revolution. Likewise, when we understand the forces on either side of a mental tug-of-war, we can tip its balance.

Lewin, who was Jewish, fled Germany for the United States in the 1930s and began working tirelessly on what he called "action research." Rather than staying in the lab, he sought out real-world problems, diagnosing the psychological forces that caused them and devising tweaks to encourage wise, healthy, or productive choices. One of his first cases was Harwood Manufacturing, a textile company that had recently moved its operations from New England to the tiny Appalachian village of Marion, Virginia. Harwood had trouble recruiting skilled factory workers to its new location, so the company hired new ones: mostly young, inexperienced women from the surrounding mountains. Supervisors trained them for twelve weeks, then pushed them to work as quickly as possible. To gin up motivation, they offered large, competitive bonuses for fast workers and heaps of criticism for slow ones.

It was a dismal failure. Virginians worked at half the rate of New Englanders and turned over twice as often. Many quit before their training even ended. Lewin realized that this reflected a toxic tug-of-war. Yes, money motivates people, but workers in Marion were already earning much more at Harwood than they would at any other job in the area. For them, bonuses were a weak psychological pull. By contrast, rushing to outpace their neighbors had real drawbacks—anxiety, fatigue, and hostility.

Lewin realized that he could realign these forces by swapping out competition for cooperation. He organized a new type of training. Instead of a company-wide focus on individual performance, groups of new Harwood employees held informal meetings and together settled on a reasonable rate of production. This system shifted their motives. Workers had chosen their target, rather than having it forced on them. Productivity meant camaraderie, not isolation. Lewin's strategy worked: Democratically organized teams not only produced more; their morale shot up as well.

Over the years, Lewin used action research on countless problems—from food choices to race relations. Generations of scientists have taken up his mantle, by making good decisions easier. Some of these techniques are called "nudges": small, subtle changes to a person's situation that inspire big changes in their actions. For instance, when a company makes retirement saving the default option for new employees, about twice as many workers save money than when they have to actively choose it. In countries where organ donation is the default, more than 80 percent of people sign up; in countries where it is not, that number is closer to 20 percent. Nudges increase college enrollment, energy conservation, voting, and vaccination rates, often much more efficiently than other policy strategies.

A growing number of psychologists have used similar approaches to help people choose empathy, even when they might avoid it. Empathy builders take Lewin's lead. As we have in this chapter, they first diagnose the forces in empathy's psychological tug-of-war. Then they alter them—amplifying empathy-positive forces, diminishing empathy-negative ones, or both.

After dismantling seminary students' empathy, Dan Batson spent the rest of his career showing he could rebuild it just as quickly. In one particularly clever study, Batson flipped compassion collapse on its head. This was the late twentieth century; the AIDS epidemic was roaring. So was the stigma surrounding its victims, who were often blamed for their illness or treated as radioactive. Thousands were suddenly afflicted, yet many Americans didn't know them. The victims were statistics, and strangers—two big reasons not to empathize.

Batson knew that people naturally care about single individuals and their stories. Could he leverage that to get them to empathize with a whole group? To test this, he played University of Kansas students a recording of Julie, a young woman living with HIV, who describes the illness's toll:

> Sometimes I feel pretty good, but in the back of my mind it's always there. Any day I could take a turn for the worse. And I know that—at least right now—there's no escape. . . . I feel like I was just starting to live, and now, instead, I'm dying.

All the students in Batson's study heard Julie, but he encouraged some of them to *listen* to her. "Imagine how the woman who is interviewed feels about what has happened and how it has affected her life," their instructions read. Unsurprisingly, this prompt increased people's empathy for Julie. But, more important, participants who imagined how Julie felt also came to care more about other people living with HIV or AIDS. They were more likely to agree with statements such as "Our society does not do enough to help people with AIDS," and reject victim-blaming claims such as

"For most people with AIDS, it is their own fault that they have AIDS."

Empathic nudges can be shockingly simple. One of the simplest—if also one of the most cynical—is to pay people for thinking about one another. My favorite research using this technique answers a question I get quite often: Are women more empathic than men? The stereotype runs deep, and women *do* exhibit more empathy than men in many studies. According to the Roddenberry hypothesis, these differences are here to stay: an eternal Venus-Mars problem. But maybe instead of being incapable of empathy, men are merely uninspired to work at it. If that's the case, the right incentives should change their behavior.

In one set of studies, men and women viewed videos of people telling emotional stories and then guessed how the speakers felt. Men performed more poorly than women. But in a follow-up, researchers told viewers that they'd be paid for accurately understanding speakers. This eliminated the empathic gender gap. A few years later, a separate research team told heterosexual men that women found "sensitive guys" attractive. Men who learned this eagerly turned up their empathy—the emotional equivalent of someone sucking in his gut as a fetching stranger walks by.

In addition to being hilarious, this second study demonstrates that incentives come in many forms. Money works, but so do sex appeal, social connection, and pride.

Other nudges seek to help us overcome tribalism. Sure, we typically care more for people in our group than for outsiders. But who counts as part of our group? People contain multitudes—you might be a woman, an Ohioan, a cellist, and an anesthesiologist. Each part of you carries with it a different definition of your group, and some of our "selves" are more inclusive than others. If I think of myself as a Stanford-ite, UC Berkeley becomes a bitter enemy; caring about its students (or football team) becomes an uphill struggle. But if I focus on being a California academic, Berkeley professors become part of my tribe—worth my time, attention, and empathy.

One clever set of studies leveraged this idea amid one of the most bitter conflicts in the known world: British soccer fandom. Psychologists recruited avid fans of Manchester United. Fans wrote about what ManU meant to them, and were told they would record a short video tribute to their team in another building. Then, in an echo of the Good Samaritan study, participants crossed paths with a jogger (actually an actor) who twisted his ankle and fell to the ground. In some cases, he wore a ManU jersey; in others, he wore the colors of Liverpool—ManU's most hated rival at the time—and in still others he wore a plain jersey. Over 90 percent of participants stopped to help fellow ManU fans, but if the jogger wore a Liverpool jersey, 70 percent of them walked right by him as he writhed in pain.

This is classic tribalism, and yet a simple nudge made it disappear. In a follow-up, the researchers asked participants to write not about ManU, but instead about why they loved soccer. Again, they set out across campus to record their videos, and again they ran across a jogger in trouble. This time, though, they helped Liverpool fans almost as often. Joggers with plain jerseys were still left behind, so a second message of this work is that if you need others' help, you're probably better off belonging to any tribe than to none.

This research demonstrates that the right psychological pull can make empathy win out. But then again, most of these studies were conducted with college students—perhaps a relatively empathic bunch already. Maybe Klansmen, criminals, and psychopaths are just meaner. Psychopaths are particularly challenging: They can figure out what others feel, but they simply don't care and so use their social savvy to manipulate and harm others. Society often writes such people off as incapable of change; even the way they're punished reflects this perspective. Psychopathic—as opposed to non-psychopathic—criminals are more likely to be killed by the state, even though it's not clear they actually reoffend more often. The thinking seems to be that there's no hope for them and that the rest of us would be better off without them.

Making psychopaths care is perhaps the hardest possible test for empathic nudges. A few years ago, the neuroscientist Christian Keysers and his colleagues traveled to prisons around the Netherlands to try it. They scanned the brains of psychopathic and non-psychopathic criminals while showing them images of people in pain. Unlike most people, psychopaths didn't show a mirroring response. That supports the fixist story: Psychopaths' lack of empathy is "hardwired" into their brains. But Keysers's team ran a second version of the study, this time taking a page out of Batson's book. They asked psychopaths to focus on victims' pain and to do their best to imagine how it felt. When psychopaths did this, their brains mirrored suffering in almost exactly the same way as non-psychopaths.

If psychopaths can turn their empathy up, surely the rest of us can, too. But do these changes really change us? Mobilists describe the mind as a muscle: Just as we can become stronger through exercise, the right practices can grow our intelligence or shift our personality. But muscle comes in more than one form. Some fibers—known as fast-twitch—are thick, powerful, and quick to tire. They allow you to sprint, squat, and lift weights, but not for long. Slow-twitch fibers are thinner and weaker but more durable, supporting extended effort, such as marathon running.

Dan Batson, Christian Keysers, and other psychologists produce fast-twitch changes in empathy. They shift people's motives, encouraging them to tune toward empathy. The effects of their prompts might last a minute or an hour, but they are unlikely to endure. Hopefully, seminary students do not constantly feel rushed and can find time to help people in need. On the other hand, most of us do not have the energy to think deeply about everyone we run into. And without being prompted to imagine someone else's situation, psychopaths will likely go on being callous. Unusual situations push people along their empathic range, but in the absence of prompts, they might slide right back to their set point.

A larger goal would be to build slow-twitch empathy: not just moving people to the right of their range, but helping them stay there. This would take more than one instance of psychological tuning—the same way that it takes more than one jog to strengthen our hearts and lungs. It requires chronic, repeated experiences, the type that spur psychological mobility. As we've seen, these sorts of tectonic shifts can happen—if you're born into a nurturing family, or if you experience great adversity—but can they be designed? Few scientists have tried. Even attempting to do so would require enormous amounts of time, labor, and money—and there's no guarantee it would work. But if we are to fight for kindness, we should first establish that the fight can be won.

A team at Germany's Max Planck Institute, led by Tania Singer, recently provided a dramatic answer to that question. Singer is one of the neuroscientists who first put brain mirroring on the map. In the early 2000s, she recruited romantic couples to her lab and scanned one of them using MRI while they and their partner took turns receiving electric shocks. Participants activated the same parts of their brain when they felt pain and when the person they loved did. Not only that, but more empathic people exhibited stronger mirroring. This convinced much of the scientific community that some people do simply care more than others, a difference that lives deep in their brains.

But Singer herself never believed that. While earning her PhD, she had researched neuroplasticity and knew that just because something occurs in the brain, that doesn't mean it's fixed. She made contact with Buddhist monks who thought of empathy as anything but hardwired. In their traditions, compassion was work, and many of them practiced it for hours every day.

For the second act of her career, Singer decided to test whether these ancient techniques could tune people's brains for kindness. She amassed more than seventy researchers and teachers for a wildly ambitious project. In a two-year period, they ran roughly three hundred participants through thirty-nine weeks of intensive training. In three-day-long retreats and daily guided practice, students honed their meditation skills. They learned to sharpen their attention and carefully notice their breathing and the sensations in their bodies.

They then trained that focus on others. In *metta,* or loving-kindness meditation, students focus on their desire to alleviate suffering and increase well-being. They wish goodwill first on themselves, then friends and family—easy empathic targets. *Metta* then requires them to stretch this warmth toward strangers, people they dislike, and eventually all living beings. Singer's group paired students up to practice empathy together. In each pair, "speakers" shared emotional stories, and "listeners" practiced *metta* toward speakers. Then they switched roles and started over. Through a smartphone app, pairs came together almost every day to practice together.

Singer and her team carefully measured students' experiences before, during, and after training. What they discovered was striking. Over time, students found it easier to pay attention for long periods—a rare skill in our information-addled age. They described their own emotions in more nuanced terms and pinpointed others' feelings more accurately. Training was the emotional equivalent of wearing glasses for the first time: The world became more vivid, full of details they never knew they'd missed. They also acted more generously and found it easier to recognize their common humanity with others, even people very different from themselves. When encountering others in pain, students felt a greater desire to help than they had before.

The changes didn't stop there. Singer and her team scanned students with MRI before and after training. They examined not just their empathic physiology—how students' brains reacted to other people—but their anatomy as well: the shape and size of their cortex. Remarkably, they found that empathy-related parts of the brain grew in size after kindness training. As we've seen, the brain changes in response to the skills we learn and the habits we pick up. But Singer's team showed for the first time that through purposeful effort, people can build long-term empathy and, in the process, change their biology.

SINGER'S PROJECT IS a proof of concept: It shows that we *can* build slow-twitch empathy—moving ourselves along our range and changing our brains. But it's hard work. Most of us would like to get in better shape but are not in the mood to run ultramarathons. The Singer program requires weekend-long retreats and daily practice for as long as a full-term pregnancy. It's Olympic training when most of us can only spare a few trips to the gym.

Is there a less demanding way to encourage empathy that lasts? One possibility comes from changing people's beliefs. Carol Dweck teaches people that they can grow—become smarter, more open-minded, and more empathic. This causes them to work harder in the moment, persevere in the face of challenges, and notice their own strength. But mindsets, for instance about intelligence, can also produce slow-twitch change by turning into self-fulfilling prophecies. People who believe in themselves do things that give them even more reason to believe. They adopt habits of mind that work over the long term.

Building on these ideas, my graduate student Erika Weisz recently set out to see whether a similar approach can encourage long-term empathy. She recruited brand-new Stanford freshmen and asked them to complete a "pen pal study." Some students read a note from a high schooler who had moved to a new state and was having trouble making friends. We asked them to write an encouraging response. In particular, we taught freshmen that empathy is a skill that people can build—using some of the same evidence you've read about—and that their high school pen pal could use this to make new connections. Other freshmen read a note from a high schooler who was having grade troubles; they were told to persuade their pen pal that intelligence is malleable.

Students asked to write about empathy got the message. One wrote, "I know it seems hard to be social, be vulnerable, and exercise empathy. It may feel like you are simply unable to connect with certain people, but studies have shown that . . . with effort and practice you can mold your empathy." Another counseled, "Empathy is like a habit or a skill we can learn and practice for improvement, much like reciting vocabulary over and over again or practicing in sports." A third closed their letter, "Remember that your ability to connect with people is entirely up to you. With just a little effort put forth, you can make friends. Go do it!"

We really did send these students' notes to younger kids (more on this later), but this study wasn't actually about changing high schoolers' minds. Research demonstrates that when people try to convince someone else of something, they often convince *themselves* in the process. Erika and I used this as a back door to changing college freshmen's mindsets, encouraging them to adopt a mobilist view of empathy.

The exercise stuck. Two months later, students who had written that empathy is a skill continued to believe it. Remarkably, it also appeared their empathy really *had* grown. They were better at decoding other people's emotions than students who had written about intelligence. And in the crucial first months of college, students who wrote about empathy as a skill reported having made more close friends than their peers.

Our results are preliminary, and they need to be confirmed by future studies. But they make a promising point. Erika's intervention was much simpler than Singer's, and required only a few hours from our participants. And yet it produced at least some lasting change. This suggests that we can build empathy pretty efficiently. People can change how they approach everyday situations—the stories they hear, the people they meet, and the technology they use. The right tweaks make caring come naturally, turning an up-hill climb into a downhill stroll.

Throughout the rest of this book, we'll explore these strategies. Like Lewin, we'll leave the lab and meet people where they are. We'll encounter individuals lost in the depths of hatred, isolation, and stress: pushed away from empathy by their own pain, by their jobs, by their phones and televisions, by the systems around them. Against the odds, they find ways to connect, building empathic habits, overcoming division, and becoming kinder people.

Their experiences point the way for the rest of us. The modern world might diminish empathy. But rather than accepting this, we can identify the forces that make it happen and push back against them.

CHAPTER 3

Hatred Versus Contact

TONY MCALEER USUALLY targeted Jews, but this time he made an exception. He and his friends in the White Aryan Resistance (WAR), dressed in Doc Martens and carrying walking canes straight out of *A Clockwork Orange*, began harassing a gay man in a park. He ran, and they gave chase through Vancouver's moonlit streets, cornering him in an empty construction site. He ducked into a long, narrow crawl space for shelter. Tony and his friends grabbed loose rocks and whipped them at him. They skipped like stones across a pond's surface, disappearing into the darkness. Every few throws, a rock hit its mark, and the man's screams echoed back at them. "It was like a game," Tony said. He felt nothing.

Tony grew up near that construction site. His father was a psychiatrist who had immigrated to Canada from Liverpool, England. He worked long hours and on most days returned home after Tony had gone to sleep. He badly missed England and built a full British pub in their basement, complete with a copper-topped bar and home-brewed beer ("He thought Canadian beer was piss"), retreating there most nights. "At the end of a day of hearing other people's problems," Tony remembers, "the last thing he wanted was to listen to ours."

The elder McAleer did, though, *create* problems. At age ten Tony walked in on his father with one of his mistresses, after which the family detonated, leaving Tony angry, confused, and displaced. His taste in music began skewing punk—from Elton John to the Clash—and his grades plummeted. Together with his parents, Tony's teacher decided to improve his academic performance by stick rather than carrot. When Tony scored below a B on a major exam or assignment his teacher caned him with a ruler. Unsurprisingly, this only deepened Tony's anger. He rebelled at every turn, earning the school record for most consecutive detentions.

Hatred blooms from a tangled, poorly understood system of roots. People who commit violence on the basis of race, religion, or gender identity are disproportionately young and male. They also tend to be economically displaced; periods of high unemployment yield increases in hate crimes. But in a recent survey, a strong thread tying hate group members together was a history of abuse—almost half of the offenders reported some form of physical or sexual violence in their past.

Tony saw this pain among other young white-power enthusiasts. "We were like an island of misfit toys. Everyone was damaged. Everyone was angry. No one would know a healthy relationship if it kissed them on the lips."

The vast majority of neglected children do not join the White Aryan Resistance. That takes a confluence of events in which a person finds acceptance in a community built on hate. For Tony, this began with a return to his ethnic home. He begged his parents to let him transfer schools, and for tenth grade they sent him to a seaside boarding school in England. This didn't quell Tony's aggression—weeks in, he organized a student uprising in his dorm simply to stir up mayhem—but it did give him something on which to hang his identity. British skinhead and Oi! bands were his catnip: He loved their driving sound and took pride in their celebration of his mother country.

Tony returned to Vancouver with a buzz cut and a pair of Doc Martens. Shortly after, two other punks approached him outside a Black Flag concert, hoping to jump him for his boots. He talked his way out of the mugging and befriended the boys, who took him deeper into the skinhead music scene. Tony ramped up to overtly racist bands such as Skrewdriver, who encouraged fans to protect the white race by any means necessary. The skinhead scene offered him two things he had sorely lacked. The first was an outlet for his aggression. At sixteen he got into his first brawl and lost badly; he didn't care. "I remember the thrill of it. The high was like scoring a goal in a championship game. It was addictive." Tony got a similar rise out of pushing social boundaries—saying taboo things about black people, or Jewish people, or gay people. It felt rebellious, confident, and exciting. By twelfth grade, he had pinned a swastika onto his camo-green jacket.

The movement also offered Tony a chance to flex his intellect. He "nerded out over National Socialism" and specialized in Holocaust denial, collecting details that—when squinted at through a warped, anti-Semitic lens—cast doubt on the standard stories. Outsiders usually imagined that skinheads would be angry and stupid. He would overwhelm them with an avalanche of factoids, waiting for them to grow frustrated so he could declare victory. "I wasn't the greatest street fighter," Tony remembers, "but I was the great debater." Sometimes, after coolly stating his case to a crowd, he'd lean close to his opponent and whisper something awful into their ear, to further break their composure.

Tony's wit won him respect, and he rose quickly as a leader of local white supremacists. He also brought Canadian bigotry into the twenty-first century. The Internet was still young, but Tony created a website for Resistance Records, the first white-power record label in North America. He also founded Canadian Liberty Net, an automated phone service people could call to hear hateful voice messages about Jews, blacks, and Native Canadians, like Moviefone for racism. At the peak of his influence, Liberty Net received hundreds of calls each day.

As Tony rose through the ranks, he grew more extreme and less humane. "I was like a frog, being boiled in a pot that gets one degree hotter at a time." He cut ties with Jewish and Asian friends from his childhood. His worldview grew virulent and paranoid. He saw his culture besieged by outsiders. Liberty Net's main message put it clearly: "White people of America are surrounded on all sides by a rising tide of the lower races who envy and hate them, who are streaming in in uncounted numbers and who are encouraged to dispossess whites by the incessant anti-white drum beats of the alien controlled media."

It's not just extremists like Tony who show these tendencies. People who feel threatened by outsiders often grow aggressive and reactionary. In two recent studies, researchers presented white Americans with evidence that minority groups would soon eclipse whites in population and reduce their economic advantages. In response, whites tacked right politically, especially turning against policies that help minorities.

Tony's utopia—a white, closed Canada in which every Jewish person "just went someplace else"—would be a nightmare for millions. But he didn't necessarily want outsiders to suffer. He just didn't care whether they did or not. "Ours was a civilized barbarism. When people think of hate, they see a red-faced, screaming person. That's hate mixed with anger. True hatred is a profound lack of connection. . . . At that time I couldn't connect with other people's pain, or with my own." Tony's hatred was less like a scream and more a cold silence. His beliefs had granted him friends, power, and status. In exchange, he grew numb. "I didn't lose my humanity," he recalls. "I traded it for acceptance and approval."

IF HATRED IS a disease, it is mutating. As our culture cures itself of one strain, new ones take its place. Over recent decades, Americans have accepted interracial and same-sex marriage, but political animosity has soared. In 1960, Americans were asked how they would feel if their child married someone of the opposite political party. Five percent of Republicans and 4 percent of Democrats said they'd be displeased. By 2010, this number had ballooned, to half of Republicans and a third of Democrats. As our ideals drift further apart, members of each party like each other less and discriminate against each other more. We also show little interest in each other's perspectives. In one recent study, both Republicans and Democrats paid money to *not* listen to the other side's opinions.

People effortlessly carve the world into insiders and outsiders. Divisions between groups can be biological (old versus young), traditional (Real Madrid versus Barcelona), momentary (one pickup basketball team versus another), or even made up. Assemble a group of strangers, give half of them blue armbands and the other half red ones, and they will build new prejudices on the fly, judging their fellow reds (or blues) as kinder, more attractive, and more capable than the sinister blues (or reds) on the other side. Boundaries between insiders and outsiders destroy virtually every type of empathy scientists can measure. When people encounter outsiders in pain, they report less empathy, feel less anxious, and imitate the person's facial expressions less than when the victim is an insider.

Ignoring outsiders' emotions makes it easier to oppress them. A century ago psychiatrists shackled psychotic patients in ice water baths for hours on end, claiming they couldn't feel cold. One physician working in the nineteenth century remarked, "What would be the cause of insupportable pain for White men, a Negro would almost disregard." Even now, people guess that a black person will feel less pain from a syringe or burn than a white person. This denial worms its way into medical treatment, where black patients receive less pain relief than their white counterparts.

Many people strip each other of humanity with a shocking lack of self-awareness. In 2015, the psychologist Nour Kteily and his colleagues showed people this rating scale:

Some people think that people can vary in how humanlike they seem. According to this view, some people seem highly evolved whereas others seem no different than lower animals. Using the sliders below, indicate how evolved you consider each of the following individuals or groups to be.

It seems unthinkable to deny that any group of people is fully "evolved." Yet in one of Kteily's studies, Americans (who in this case were mostly white) rated Arabs at only about 75 percent evolved, and Mexican immigrants as about 80 percent. Those who rated Muslims as less evolved were also more likely to support anti-Muslim immigration policies or the torture of Muslim detainees. During the 2016 Republican primaries, those who viewed Mexican immigrants as less evolved also endorsed statements by candidate Donald Trump such as "People are coming from all over that are killers and rapists."

Dehumanization silences empathy at the most basic levels. Imagine you could watch a person's brain in real time while someone else receives painful electric shocks in front of them. Within a fraction of a second, you could tell whether the victim and observer were part of the same group. If they were, the observer would produce neural mirroring; if not, that mirroring would be blunted or nonexistent.

Conflict worsens the situation. Sports rivalries, ethnic clashes, and everything in between flip empathy on its head. The psychologist Mina Cikara studies "schadenfreude," or enjoyment of others' pain. She's found that Red Sox and Yankees fans activate parts of their brain associated with reward when watching their rival team lose, and that people smile when imagining misfortunes befalling outsiders they dislike.

This work might seem to imply that we can't help but empathize with insiders and *not* care about outsiders, dooming us all to some degree of prejudice. For extremists like Tony, empathy would be forever out of reach.

At twenty years old, Tony would have agreed. His life was defined by hatred. He appeared on *The Montel Williams Show* as an avatar of white supremacy. He figured he'd be dead or in jail within a decade. But over the next several years, he met three people who changed those odds. The first two were his children. At twenty-three, Tony had a daughter; at twenty-four, a son. On the outside, life remained tumultuous. The Canadian Human Rights Council brought a lawsuit against Liberty Net and ordered Tony to appear for a hearing. He sparred with lawyers all morning, and during lunch ran the six blocks to the hospital to witness his son's birth. Soon after, an ugly breakup left Tony a single father of two.

He decided to avoid following in his father's footsteps. "Instead of being the type of dad I had, I tried to be the kind of dad I would have wanted." Tony doted on his kids, and in exchange they offered him connection he hadn't felt in decades. "It's safe to love a child. They're not capable of rejection, or shame, or ridicule." Being an active single father also cast Tony in a new light. "I got lots of praise heaped on me. It's not fair—if I was a woman I wouldn't have gotten that. But I enjoyed it." This was 180 degrees from the villain Tony was used to playing. A stranger who might have punched him for his beliefs instead patted him on the back for raising his kids. This gave Tony a chance to see himself differently.

Kids were also expensive, and Tony worried that his public bigotry would make him unemployable. He decided it was time to leave the skinhead movement, and "went dark," steadily lowering his profile. He put his tech savvy to use as a financial consultant for Internet start-ups. He still loved raucous parties, but he swapped his old Aryan punk shows for Vancouver's rave scene. On weekends, he dropped his kids off with their grandparents and embarked on twenty-four-hour "escapes," fueled by electronic music and tablets of ecstasy. His old friends had moshed and brawled; his new ones swayed and hugged. "It was the polar opposite of what I'd been into before." On some nights he would return home, still high, and listen to Skrewdriver's hard-charging white-power anthems, filled with melancholy.

Fatherhood mellowed Tony, but it didn't change his beliefs. He saw caring for his children as his best shot at supporting the white race: a bigoted version of "Think global, act local." Yet his animus toward black, gay, and foreign people mattered less to him. "The ideas were still in my head, the questions were still in my head . . . but it was like, 'So what? Look at what my kids are up to, they're fantastic.' "

Tony's most stubborn prejudice was against Jews. That domino fell after he met a third person. To better himself, he took classes on everything from public speaking to mindfulness. One of his teachers was Dov Baron, a leadership trainer. Tony and Dov shared British roots, bonded over their shared love of Monty Python, and became fast friends. Dov offered one-on-one counseling, and a mutual friend bought a session for Tony. Midway through their conversation, Tony apprehensively copped to his skinhead past. Dov smiled and said, "You know I'm Jewish, right?" Tony was mortified, but Dov comforted him. "That's what you did, but not who you are," he told Tony. "I see *you*."

Tony spent the next half hour crying in Dov's office. "Here was this man who loved me and wanted to heal me, and here was I, a person who had once advocated for the annihilation of his people." Tony felt he didn't deserve a shred of compassion from Dov, but Dov extended it nonetheless. This cracked Tony open. He'd created a surface of hatred to cover his shame and loneliness. Once someone accepted him warts and all, he no longer needed it.

Tony began exorcising his past, opening up about his history, and accepting responsibility for the pain he'd caused. He feared that his clients would abandon him after learning what he'd done, but only a few did. At a party, he weepily described his gay bashing to a group of gay men; one cursed him and walked away, two became close friends. In Tony's first ever act of hatred, he had vandalized a Vancouver synagogue. Recently, he went back there to confess and listen. Over and over again, he saw shades of Dov: People did not minimize what he had done, yet they were willing to see Tony as more than his past.

A few years ago, Tony traveled to a Holocaust museum. In the past, he would have taken to its exhibits "like a lion looking to pounce" on any fact he could dispute. Instead, he lingered for hours, looking at pictures of the dead, reading each of their notes, examining each of their mementos. That night in his hotel room, he lay down and felt something heavy on top of him, like one of those lead X-ray aprons. "I could feel it moving up my chest and into my throat, and boom, out it came: the insight that my denial of their pain was a denial of my own." He wept through the night, filled with a feeling he'd long kept at bay.

HATRED BURIES EMPATHY but does not kill it. Tony's conversion highlights a powerful way to get it back.

In 1943, a race riot seized Detroit. World War II had transformed the city into a weapons factory, and people poured in from around the country as manufacturing boomed. Housing became scarce. Black workers were excluded from housing projects and often paid triple the rent whites did. When the city earmarked a new project for black tenants only, whites burned crosses outside it. As summer arrived, racial tension boiled over. On June 20, blacks heard that a white mob had thrown a woman and her child off the Belle Isle Bridge; whites heard that blacks had raped and killed a woman on the same bridge. Neither event had actually occurred, but the imaginary mobs spawned real ones. They clashed, and in the next thirty-six hours, thirty-four people died, hundreds were injured, and thousands were arrested.

It was a national disgrace and a low point for American race relations. But there was a glimmer of hope: whites and blacks who had worked or studied with members of the other race were far less likely to join in the riots, and more likely to engage in peaceful behaviors, such as sheltering other-race individuals from violence.

The psychologist Gordon Allport noticed this, and saw a trend: The better people knew outsiders, the less they hated them. This was true elsewhere as well. Seventy-five percent of residents in all-white housing projects said they would dislike living alongside blacks, but only 25 percent of white residents in mixed projects *actually* disliked having black neighbors. Sixty-two percent of soldiers in all-white platoons opposed integrating the armed forces; among whites who had been in a mixed platoon, that number was 7 percent.

In his magnum opus, *The Nature of Prejudice,* Allport reasoned that bigotry often boils down to a lack of acquaintance. Its antidote was just as simple: Bring people together, and they'll awaken to their common humanity. A similar thought led Mark Twain to quip, "Travel is fatal to prejudice, bigotry, and narrow-mindedness, and many of our people need it sorely on these accounts." In psychology, this idea came to be known as "contact theory," and it caught fire. Allport's book, published in 1954, became a bestseller; he delighted in spotting it at airports and malls alongside beach novels. Thanks to him, optimists everywhere came to believe that hatred was a misunderstanding and that contact could fix it.

Allport stressed that contact would not always work. In some cases it could make things worse—for instance, whites who merely *saw* more blacks, but didn't get to know them, might perceive them as a threat. Time has proven Allport right. The visibility of immigrants in the United Kingdom fueled a nationalist wave that crested with Brexit. In Canada, Tony used the presence of immigrants to whip up white aggression. Even moderate people can be driven toward prejudice by the wrong type of contact. In a recent study, the political scientist Ryan Enos planted Latino passengers on a Boston commuter train at the same time each morning for ten days. White commuters who had been on a train with Latinos grew less tolerant of immigration than they had been before, or than passengers who took the very next train.

Even when contact doesn't hurt, it might not help. "Goodwill contact without concrete goals accomplishes nothing," Allport wrote. He laid out a recipe for how to make it useful: Bring groups together and give them equal status, even if one group has more power the rest of the time. Focus on their mutual goals. Make it personal; let people learn about each other's idiosyncrasies. And support cooperation between groups through the institutions around them. Fulfill these tenets, Allport claimed, and contact could do wonders.

The theory might sound naïve—less rigorous science than Haight-Ashbury handholding. But it's one of the most well-studied concepts in psychology. In a recent analysis of more than a quarter of a million people, the pattern was clear: The more time someone spends with outsiders, the less prejudice they express. Contact warms sentiments toward many types of outsiders. Imagine two straight, young, able white people born in the United States. Evidence suggests that the one who gets to know a diverse group will exhibit less bias toward black and Hispanic people, immigrants from Asia, Mexico, and Central America, elderly and disabled people, and the LGBTQ community than their more sheltered counterpart.

Contact can work even when people don't seek it out. White college freshmen randomly assigned to black roommates are less prejudiced by spring than students with same-race roommates. It doesn't take a school year to help people warm up to new groups, either. In one recent study, trans- and cisgender canvassers went door-to-door in Florida to discuss transgender rights. After meaningful conversations with trans canvassers, residents' transphobia dropped substantially, and they remained more tolerant three months later.

THE PUNCH LINE is simple: Hatred of outsiders is ancient but not inevitable. When people work, live, or play alongside each other, divisions between them melt.

To understand why, remember that empathy is a choice, and conflict gives people great reasons to avoid it. When groups compete for scarce resources, they must circle the wagons and scrap for their side. As Tony puts it, "Multiculturalism and diversity are great when everyone's doing well, but when you're fighting your neighbor for a crust of bread in the street, all bets are off." Tribalism becomes natural and—from evolution's perspective—wise. A linebacker who feels the pain of a running back would have a hard time doing his job; a soldier would find hers impossible. As a result, people in conflict don't merely misplace their care; they actively throw it away. In one series of studies, conservative Israelis reported that they would prefer not to empathize with Palestinians. This preference, in turn, predicted their *actual* lack of empathy, for instance when they read about a Palestinian child with cerebral palsy.

Even if callousness is a smart choice during war, it's a terrible way to achieve peace. Contact remedies this by giving people reasons to care about outsiders. We crave connection and will work to keep social bonds strong. When an outsider joins the ranks of our friends or colleagues, empathizing with them aligns with that goal. The benefits compound: Empathy for one outsider can lead to caring for their entire group, as Dan Batson demonstrated in his study of AIDS victims. Contact also makes avoiding empathy harder. The sorrow and hope of a neighbor, friend, or colleague are often impossible to block out.

Contact can build empathy even in the toughest settings. After sectarian violence in Northern Ireland, Catholics and Protestants dehumanized each other, but they did so less if they had friends on the other side of the conflict. White Americans who work or live with blacks or Muslims express higher empathy when members of these groups are profiled by law enforcement. Empathy, in turn, promotes solidarity. After the conflict in Ireland, people who felt empathy toward outsiders were more willing to forgive them; in the United States, white Americans who empathized with minority individuals mistreated by police were more likely to join Black Lives Matter protests.

For decades, scientists and practitioners have tried to bottle the lightning of contact. In Hungary, the Living Library School Project offers people the chance to converse with living "books"— individuals from marginalized groups such as the Roma who have agreed to share their stories. The Parents Circle brings together Palestinians and Israelis who have lost family members to the conflict, in hopes that their common grief can overcome their differences. And Seeds of Peace brings teenagers from Palestine and Israel together in a two-week summer camp in Maine. Students are split into "color war" teams that cut across ethnic lines. Teammates bunk together and compete with other teams throughout the camp. By focusing adolescents on their new team identity, Seeds of Peace points them away from older divisions. Even months later, campers report warmer attitudes than non-campers toward people on the other side of the conflict.

Recently, psychologists examined about seventy contact-based programs like these. Many succeed in building care and camaraderie between groups. At least some of these benefits lasted up to a year afterward. But as Allport realized, contact doesn't always work. And when it does, it's not always clear why. To use it effectively, psychologists must isolate its active ingredients. Allport's rules of engagement are a great start, but they are in sore need of updating.

Emile Bruneau is leading a new charge to sharpen the science of contact. He always yearned to understand how other people saw the world, in part because he struggled to understand his own mother. Shortly after Emile's birth, Linda Bruneau began hearing taunting, menacing voices: in the sound of a plane flying overhead, or coming from the television, or from nowhere at all. To Linda, these voices were as loud, clear, and real as any person's. As Emile grew, Linda descended further into schizophrenia.

Emile entered neuroscience in the hopes of learning more about his mother's mind. Early on, he encountered a study that astonished him. Neuroscientists had scanned the brains of people with schizophrenia. Whenever the person in the scanner heard voices, they pressed a button, allowing researchers to chart where hallucinations originated in the brain. They found that imagined voices activated the same brain areas that process sound. Biologically, they appeared indistinguishable from real hearing. To Emile, this was redemptive. While he was growing up, schizophrenia was blamed on patients' families, a view that had torn his apart. But here was a different perspective entirely. "I realized it was a biological condition . . . and something that's biological is much more tractable. . . . You think, 'This is terrible,' but there's something you can do."

Emile also traveled widely, often ending up in places riven by violence. He spent months in South Africa shortly after the fall of apartheid. He visited two journalist friends in Sri Lanka, landing hours before the Tamil Tigers attacked Colombo. Turmoil in each of these places was unique, but it also shared common themes. Most important, it warped otherwise good people. In South Africa, Emile got lost during a biking expedition and emerged from the forest hungry and bruised. An elderly woman nursed him back to health, asking nothing in return. Then apartheid came up, "and racist shit started pouring out of her mouth." It was as though she had splintered into two selves.

To Emile, conflict looked a lot like schizophrenia: stranding people in versions of the world that are real to them, but not to others. He began to suspect that clashes between groups might attack the brain like a psychiatric disorder. And if it was biological, it should be treatable. He set out to examine current treatments, traveling to Belfast to volunteer for a contact-based program that brought Catholic and Protestant boys together for three weeks. "All of them bunked together in an enormous gymnasium, spending their days designing murals, playing music together."

"It was a colossal failure." The boys were friendly enough toward each other during the three weeks, but on the last day two kids got in a fistfight, which quickly exploded into an all-out brawl between Catholics and Protestants. Students who had played together an hour before snapped back into their old identities in seconds. As Emile broke up a fight, he heard one boy scream at another, "You orange bastard!" It dawned on him that the boy was referring to William of Orange. "The epithets they were throwing at each other were six hundred years old. I thought, 'Holy shit, this is deep.'"

He also realized that contact programs tended to throw the kitchen sink at conflict. They clumped together dozens of activities and discussions. Emile felt that a more precise approach could identify exactly which pieces of contact helped the most, when, and in what ways. "What are the primary ingredients that allow a program to work? How do they interact with each other? Which interventions work best for which type of people?" These questions sound simple, but the research didn't address them.

Emile decided to try to answer these questions himself. Over the years, he's explored the corrosive effects of conflict on empathy, and teamed with peace-building organizations to probe how and when contact works. Emile doesn't reinvent the wheel; his partners know their conflicts better than he ever will. He takes their template, tinkers with the materials they produce, and tests which version works best.

Sometimes, Emile's answers contradict conventional wisdom.

Gordon Allport believed that contact was most effective when groups came together under equal status, even if one group was richer or more powerful the rest of the time. Most conflict resolution programs stick to this principle, for instance ensuring that Israelis and Palestinians are given similar airtime during discussions. Each side is encouraged to listen closely and take the other's perspective.

People from majority or high-power groups often walk away from these sit-downs with a warmer view of the other side. Minority or low-power individuals, though, often don't. They *already* understand the majority's perspective, because they have to in order to survive. In a recent interview, the comedian Sarah Silverman summed up this feeling. "Women are so keenly aware of the male experience because our entire existence had to be kind of through that lens. Whereas men have never had to understand the female experience in order to exist in the world."

Minority individuals, Emile thought, might be weary of perspective taking. Instead of fetishizing equality, contact programs could respond by promoting balance. If one group is silenced the rest of the time, perhaps they should be given *greater* status when the groups come together, a chance to be heard by the more powerful side. Instead of perspective taking, they might benefit from "perspective giving." To test this idea, Emile set up shop in a public library in Phoenix and paired Mexican immigrants and white U.S. citizens who had never met. In each pair, one person was assigned to the role of "sender" and wrote a short essay about the hardships facing their group. The second person, the "responder," read and summarized the essay, and passed their reflections back to the sender. Each person then described how they felt about their partner and the ethnic group they represented.

White Americans responded to contact just as Allport would have predicted: After playing the role of the responder, they felt better about Mexican immigrants. Mexican immigrants, though, felt *worse* about white Americans after listening to the complaints of this richer, more powerful group. They felt better about whites after playing the role of sender. Emile reran the study in Ramallah and Tel Aviv, setting up video chats between Palestinians and Israelis. Israelis—like white Americans—felt best about Palestinians after hearing their stories. Palestinians, though, felt best about Israelis after telling their own stories and having an Israeli listen to them. Contact worked best when it reversed the existing power structure, rather than ignored it.

Emile has dissected hatred around the world, but lately he has focused on home, and the growing white nationalist movement in the United States. The "alt-right" has become increasingly emboldened, and more openly hateful. In August 2017, they assembled with neo-Nazis in Charlottesville, Virginia, to protest the removal of a statue commemorating Robert E. Lee. The demonstration turned violent, and an alt-right activist drove his car through a crowd of counterprotestors, injuring many and killing counterprotestor Heather Heyer. The scene looked like the West Bank, not an American college town.

At the height of his WAR period, Tony McAleer's empathy for outsiders atrophied. Today's white nationalists follow suit. They dehumanize outsiders, rating Muslims as only about 55 percent evolved on Nour Kteily's scale. They exhibit blunted responses to others' emotions and find violence a reasonable means for advancing their beliefs. It's easy to fear alt-right proponents and even easier to write them off as hopeless bigots. But Tony's story demonstrates that lost souls can still reclaim their humanity. Can we engineer circumstances that will help them?

AFTER RENOUNCING WHITE supremacy, Tony discovered an online journal titled *Life After Hate,* full of stories like his own. People chronicled their experiences with hate groups and how they got out. Tony became an active contributor. In 2011, he and other contributors were invited to an extraordinary meeting. Google Ideas (now Jigsaw, a division of Alphabet) brought together about fifty "formers"—ex–hate group members—to discuss strategies for preventing extremism. "It was nuts," Tony remembers. "You had IRA members sitting across from Jihadists and neo-Nazis. These are people who would have tried to kill each other before."

Despite their obvious differences, people at the meeting shared common stories. Many had used hatred to cover wounds from their childhoods. Many had escaped it after finding new meaning, especially through parenthood and friendships with forgiving outsiders. "Over and over, same reasons in, same reasons out." Tony realized that his struggles were not unique. And that meant he could guess what might help other people find a way out of hatred, too.

Along with his colleagues, Tony expanded Life After Hate into a nonprofit that now works to extract people from the dark place he once inhabited. "We stumbled through the wilderness and managed to get to the other side," he says. "From there we want to go back and help people who are where we were." It's a textbook case of altruism born of suffering.

Life After Hate infiltrates Aryan, neo-Nazi, and KKK message boards and social media pages, reminding visitors that they still have options. Hate group members and their families reach out to them often; the week after Charlottesville, they received about a hundred calls. Tony and his colleagues connect them to counselors, tattoo-removal services, and a more hopeful future.

ONE CLOUDY JULY day, Tony, Emile, Nour Kteily, and I all gathered at Northwestern University for a day-long brainstorming session. Life After Hate wanted to learn more about empathy from the psychologists who study it, and we psychologists wanted to hear more directly from people like Tony. We all wanted to find solutions. It was an unusual meeting. First, Life After Hate members told their stories.

Tony began. He had a soft, open face and wore a pink striped shirt, a shark-tooth necklace, and a bracelet made of spherical wooden beads. He looked a bit like a soccer hooligan turned folk musician.

Angela King followed him. She was a victim of intense bullying in school and, at some point, decided that the best way to stop being a target was to become the bully. She grew homophobic and racist and started committing hate crimes. She was arrested for the armed robbery of a Jewish-owned store and sent to prison. Her arms, legs, and chest were covered with swastikas; SIEG HEIL was tattooed on the inside of her lower lip. She expected prison to be a race war. To her shock, the first inmates to take her in were not fellow Aryans, but a group of Jamaican women. Over games of cribbage, they challenged her beliefs but also accepted her as a person. As she has recalled, "Aggression and anger and violence . . . were my reactions to anything and everything in my life. So, when I was treated with kindness and compassion, it was like being disarmed."

Next was Sammy Rangel, a soft-spoken man with cropped black hair who looks like he might live at the gym. He has a similar story: abuse, then hatred, then redemption through contact with an understanding outsider.

Emile, Nour, and I then shared research on hatred and how to address it. We focused on nudges that make it easier for people to get to know outsiders and harder to stereotype them. We bounced strategies around the room for a while, but the Life After Hate members pushed back. "You're trying to problem-solve this," Sammy objected, "but in these interventions, you're not a problem solver. That's a trap." Hate group members *expect* people to try to change their minds. In preparation they construct what Tony calls a "fortress of reason": protection against any argument, through counterpoints, rhetorical tricks, or plain old threats.

To get past those defenses, Life After Hate begins in a different place. "The goal can't be to change a person right away," explained Sammy. "You first need to show genuine interest in them, listen to what they have to say, and then maybe after a while find something to hold on to." Sammy here sounded like the legendary therapist Carl Rogers. Rogers felt that a psychologist's most important job is to truly listen to their patient, full of curiosity and free of judgment. Hate group members are ready to be cast away by anyone who doesn't agree with them, which is pretty much everyone else. Sammy, Tony, and Angela all once believed that everyone else *should* hate them. To puncture that shame, someone had to show them genuine empathy.

Tony made clear that empathizing with a hate group member is not the same as validating their beliefs. "You absolutely judge the ideology, the hatred, but you don't judge the person." Even that sounds like a tall order. Why spend any energy validating someone who's covered in genocidal tattoos? No one is obligated to return hate with love. Tony's friend Dov Baron and the Jamaican women who mentored Angela certainly weren't. But in accepting them, these outsiders gave Tony and Angela a chance to feel compassion for themselves. That compassion washed away the anger that had sprung from their shame.

Emile, Nour, and I believed that contact was a matter of changing people's minds about outsiders. But people who had been in the trenches were telling us something else: Contact had changed how they viewed *themselves*. For the past fifteen years, psychologists have been studying "self-compassion," people's willingness to take a kind, forgiving attitude toward their own foibles. Self-compassion and empathy toward others might seem like two sides of the same coin, but they're only weakly related, and sometimes not at all. A narcissist might forgive herself, but not others; a depressed person might forgive others, but not himself.

People who lack self-compassion often become rigid during conflict, for instance refusing to compromise during disagreements. Sammy's, Angela's, and Tony's childhoods had exhausted their self-compassion, but contact brought it back. Does their experience translate to others? There is little research on this, but one recent study found that Israeli children trained in self-compassion exhibited less prejudice toward Palestinians. Nour, Emile, and I are now designing work to examine the role of contact in building self-compassion, inspired by Sammy's, Angela's, and Tony's insights.

That day, I also realized how powerful contact with people like Tony must be for hate group members, by showing them that another life is possible. Imagine a member of the Latin Kings, or KKK, or WAR, who begins feeling doubt and calls Life After Hate. They meet Sammy. Sammy was almost killed by his mother and fled to live on the streets at eleven years old. He taped knives to his hands during a prison melee. He was once strapped down in solitary confinement and fed by a string held over his head like a fishing pole. The state of Illinois defined him as "incorrigible," beyond reform. He is now a social worker and doctoral candidate.

Psychologists usually think of contact as involving at least two people, but Life After Hate suggests that encountering our own past and future selves can be just as powerful. A thirty-year-old cringes at her sixth-grade self's embarrassing faux pas. She imagines her sixty-five-year-old self as weary and gray, but hopefully accomplished and satisfied. Both of these people are strangers to her. Hate group members might feel especially estranged from their future. As a young man, Tony didn't think he had one.

Research shows that people who can vividly imagine their future self behave more wisely. In one study, researchers scanned subjects' brains while they answered questions—Would they rather go grocery shopping or do laundry? Would they like to run a 5K next week?—and then again while they considered how their future self would answer the same questions. Most people activated different parts of the brain when imagining their current versus future selves, suggesting they saw their futures selves as someone else altogether. Others' brain activity suggested they had a stronger connection with their future selves. These people took better care of their futures, for instance, by investing smartly. In another study, individuals shown digitally aged images of their faces were likely to save more for retirement. Making contact with their future selves convinced people to treat them more kindly.

By meeting someone like Sammy, hate group members can connect to a future they might never have imagined, in which they can care and be cared for again. Stories like his also remind us that people can change, even after a lifetime of loss and alienation. Toward the end of our meeting, Sammy talked about the term "formers," which they had adopted at the Google Ideas meeting. "We call each other 'formers,' because we were all formerly in hate groups. But we also use it because everyone is forming all the time, into someone new."

Conflict and hate can sap our imagination. In George Orwell's *1984*, members of each political faction come to believe that they have always been at war. Politics, race, and identity in America feel that way now. Millions of us probably can't imagine a world where we empathize across all of these divisions. Like a camper invoking William of Orange to cuss out another, we accept that our history is an unending clash between groups, so our future must be as well.

In both cases we're wrong, and simply remembering that can pave the way for greater peace. In one recent study, Carol Dweck and others drew on her mindset research to convince Israelis and Palestinians that just like individuals, groups are capable of change. They were reminded, for instance, of the Arab Spring and the formation of the European Union. Even six months later, both Israelis and Palestinians felt more positively toward the other side, more hopeful about the possibility of peace, and more inclined to make concessions in the service of that peace. A belief in change also makes contact more fruitful—for instance, increasing people's willingness to cooperate with outsiders.

People who imagine a stronger, more connected version of their tribe might be inspired to make it a reality. When it works best, contact offers evidence for outsiders' worth and helps us believe in our own. It might also allow us to imagine a future in which outsiders are no longer outsiders at all.

CHAPTER 4

The Stories We Tell

Listen: Billy Pilgrim has come unstuck in time. Billy has gone to sleep a senile widower and awakened on his wedding day. He has walked through a door in 1955 and come out another one in 1941. . . . He has seen his birth and death many times, he says, and pays random visits to all the events in between.

—KURT VONNEGUT, *SLAUGHTERHOUSE-FIVE*

HOW MANY WINDOWS were on the front of your childhood home? What will your car sound like tomorrow morning when you start it? The last time you kicked a ball, how did it feel against your foot?

To answer these questions, you unstick yourself from time. Your body sits on a couch or at a desk, but your mind untethers, like a balloon floating to wherever and whenever you imagine. You are in front of the house where you grew up, marveling at how popular vinyl siding was back then. Or you're backing out of your garage, happy to have a coffee in your cup holder. Or you're at the park near your middle school, playing pickup soccer.

These mental voyages drag your senses along for the ride. When you envision your childhood home, your brain responds as though you're actually seeing it; when you remember the feel of a soccer ball, your brain responds as if to touch. Imagine a sunny day and your pupils constrict, as though you'd just walked out of a movie theater into the afternoon light.

Untethering can happen involuntarily. Like Billy Pilgrim, we can end up on an upsetting mental roller coaster. People with schizophrenia often don't know whether their experience reflects the real world or their imagination, while those with PTSD are involuntarily swept back into their worst moments. Depression strands us with past regrets, and anxiety obsesses us with what could go wrong in the future. Even outside of mental illness, untethering can be a burden. In one study, researchers pinged subjects' smartphones at random intervals, asking what was on their mind and how they felt. They reported being less happy when thinking about the past or future than when anchored in the present moment.

Other times, instead of being dragged into the past or future, we send ourselves there. Used this way, untethering becomes a powerful tool. It allows us to anticipate mistakes we haven't made and navigate situations we've never faced. How should you ambush that herd of bison? What should you say at that job interview? These are not questions you want to ask yourself for the first time when face-to-face with a giant mammal or a stern potential boss. By untethering, you can run simulations of these moments so that by the time they arrive, you're ready.

Untethering also solves a vexing puzzle in neuroscience. Researchers once thought of the brain as mostly reactive, responding to the world outside. Patches of neurons that control vision and reading "switch on" when you open a novel, not when you close your eyes. But around the turn of the twenty-first century, scientists discovered a set of brain regions that are most active when people do nothing. That was mysterious. Nature is ruthlessly efficient, and the brain consumes a lot of energy; why waste it sitting around?

As it turns out, sitting around is one of our most important jobs. Idle hands find trouble, but idle brains are free to daydream: planning ahead, reminiscing, and imagining. The brain's mystery regions are a steering system for untethering. Instead of reacting to the world outside, they project us into the past, future, and even alternate realities.

These same parts of the brain also play a crucial role in empathy. This makes sense, because empathy *is* a kind of untethering. When you imagine what your mom thought of the last email you sent her, or what victims of a recent mass shooting might be feeling, you take a mental trip into their world. The more you engage the brain's untethering system, the deeper you can go, and the better you understand what other people think or feel.

This is true even when those people are made up. Untethering is at the heart of a strange, venerable human pastime. Since people first cozied up around a fire together, we've told stories: first out loud, now on paper and screens. Surrounded by real people, we spend much of our free time pretending that people who never existed experienced things that never happened. Recently, psychologists have begun telling a new story about stories. More than a diversion, narrative arts are an ancient technology: performance-enhancing drugs for untethering. Stories helped our ancestors imagine other lives, plan for possible futures, and agree on cultural codes. In the modern world, they help in a new way: flattening our empathic landscape, making distant others feel less distant and caring for them less difficult.

"AND IF YOU heard the crow, you also knew about the sugar tarts!" Alice interrogates the cook as the duchess looks on. A call from the front row interrupts her.

"No, you're more inquisitive, not so accusatory."

Alice—actually a thirteen-year-old named Orrie—deftly adjusts her tone and delivers the line again. Instead of the traditional blue dress, she sports an oversize sweater and Chuck Taylors. Orrie is playing her first title role at the Young Performers Theatre (YPT). For more than thirty years, YPT has trained four- to seventeen-year-olds in acting and writing for the stage. *Alice in Wonderland* is their latest production, and today is their first day of blocking. The action onstage is choppy. The actors look more confident than any middle-schoolers I remember, but they still stumble over lines and stand uncomfortably in front of one another. The smallest actor, years younger than her peers, sticks close to an older girl, like a pilot fish swimming beside a sea turtle.

YPT's artistic director, Stephanie Holmes, might have just been transported from London's Globe Theatre: She has a soft British accent, a curly red bob, and effortless command over the group. She hops up from her seat every minute or so, with gentle, staccato instruction. Her corrections come in two flavors. First, she urges her actors to consider what the audience sees and knows. She adjusts the smallest actor to a more prominent position onstage. "Honey, your job is to make sure you're never standing anywhere the audience can't see you." She clarifies the difference between a sneeze and a stage sneeze, "not just a 'ch,' but a 'chOOO!' And throw your head forward."

Second, she asks actors to consider their character's inner life. In one scene, the king, played by a slight boy with thick blond hair, introduces Alice to "the Queen—my wife." He reads the line plainly, and Stephanie steps in. "Every time you say 'my wife,' it's because she babies you so much you have to remind yourself, 'She's my wife.' You pause because it surprises you."

Orrie is a veteran performer. She has the nervous energy—and painful-looking braces—of a typical thirteen-year-old, but even in one-on-one conversation, her voice projects. As a young child, she staged dollhouse productions of *Swan Lake* and *Sleeping Beauty* in her room. Not all the actors here are naturals. Ella—the "Dormouse," in YPT's production of *Alice*—was painfully shy as a young child. In seventh grade, she transferred to a new school and didn't speak to anyone for two weeks. She joined YPT only because her brother and Stephanie's son are close friends. Ella's early roles suited her timidity—her first credit was "Silent Cat"—but she eventually attempted more ambitious parts. As she did, she noticed herself growing bolder and more confident offstage as well. "Now I'm that super loud, chatty person at school who won't shut up!"

Stephanie teaches her actors to recognize common threads between their own experiences and those of their characters. To convey Alice's curiosity and bewilderment, Orrie channels a recent trip to Italy, her first time wandering through a foreign country. Ella recently played Belle in *Beauty and the Beast*. During a scene in which the Beast had gone missing, she was supposed to feel isolated and afraid, but she didn't know where to begin. Stephanie encouraged her to imagine how she'd feel if she were taken away from her family. By the end of the monologue, Ella wept.

Konstantin Stanislavsky, a father of the "method" acting school, described an actor's craft as the "art of experience." He trained actors to meditate on their characters' motives, beliefs, and history. When done right, he believed, this preparation conjures an authentic inner world, which bubbles to the surface in a true and deep performance. "[The actor's] job is not to present merely the external life of his character," he wrote. "He must fit his own human qualities to the life of this other person, and pour into it all of his soul."

This sounds like empathy's extreme sport, and method actors often bleed to take on their characters' lives. In *The Pianist*, Adrien Brody plays Władysław Szpilman, a musician and composer who hid out in the Warsaw Ghetto for years during the Holocaust, enduring starvation and loneliness while seeking comfort through his art. To prepare for the role, Brody ended his relationship, disconnected his phones, and moved to Europe. He spent months alone, playing the piano for hours each day and eating so little that he shed forty pounds. As Brody put it, he tried to "encourage loneliness and to encourage loss."

Actors ride their imaginations into others' minds over and over again. If that trains them in untethering, it should also enhance their empathy. Does it?

Thalia Goldstein was born to answer that question. At five years old, she choreographed a performance of "Baa Baa Black Sheep" and conscripted her two-year-old sister into the cast. They performed for their parents and grandparents, but the younger Goldstein interrupted the show by announcing she had to pee. Thalia allowed it, but froze in position until her sister returned, and then jumped back into song, picking up at the line where they'd left off.

By high school, Goldstein was a full-fledged drama geek. Her parents "firmly suggested" that she not study only theater in college, so she double-majored in psychology. After graduation, she worked in New York City as a waitress, nanny, and gym receptionist, while taking every audition she could. There were some breaks—she played Becky Thatcher in a national tour of *The Adventures of Tom Sawyer*—but other gigs were less rewarding. She spent months in rural Pennsylvania doing dinner theater, feeling lonesome and out of place. Tryouts wore on her, too. "When you're twenty-three and a woman and trying out for musical theater, the majority of the time you're being treated like an idiot. You've got to be cute and have big eyes, and that's fine so long as there's something else."

For Goldstein, there wasn't. After a near miss for a coveted part, she spent a day crying in bed, and then decided to make a change. She returned to psychology, but she wanted to keep her love of the arts alive. "I applied to do a PhD with anybody who would let me study creativity, art, or imagination." She landed in a childhood development lab. At the time, a hot topic among researchers was children's "cognitive empathy"—in essence, their ability to read minds. Most two-year-olds are oblivious to how other people see the world, but most four-years-olds are not. When exactly do kids learn mind reading, and why are some kids better at it than others?

Goldstein found these questions familiar in a different way. "Understanding someone's beliefs, desires, and emotions—that sounds like acting." Seeing a chance to combine her interests, she decided to devote her first term paper to all the research that combined cognitive empathy and theater. Not a single study ever had. Goldstein realized she had work to do.

There were hints that theater might help people understand one another. In one study, psychologists measured four-year-olds' penchant for fantasy. Did they have an imaginary friend? How often did they pretend to be an animal, or a plane, or a different person? Researchers then ran these children through mind-reading tests. In one, a child opens a crayon box but surprisingly finds a toy horse inside it. They're then asked what another child who hadn't opened the box would think it contained. A subject passes the test if they realize that even though *they* know what is in the box, a naïve kid will not. Children with active imaginations outperformed less fantasy-prone kids.

Theater flexes kids' imaginations, so Goldstein reasoned it should enhance their cognitive empathy as well. She partnered with a local arts high school, comparing drama majors to students in music and visual arts. Both groups took two empathy tests: picking out emotions in pictures of people's eyes and films of people interacting. In both cases, actors excelled. As Goldstein wrote, "Because actors 'turn into' other people over and over, they may be mind reading experts *par excellence.*"

This study alone didn't mean that acting makes people more empathic. Theater might simply attract kids who excel at empathy—a self-selecting group, already good at understanding others. Goldstein had performed alongside gifted natural actors who effortlessly entered character. She didn't think that their level of untethering could be taught, but she still believed training could make a difference. "If I find someone named Sarah Williams and teach her some tennis, there's no way she'll be as good as Venus or Serena. But she'll be better than she was before." Even if theater kids start out empathic, training might make them more so.

Goldstein devised a second, more stringent study. She tested visual arts and drama students' empathy twice: both before and after a year of schooling. Students who trained in acting did start out slightly more empathic than visual arts students. But theater training also *grew* students' cognitive empathy over the year, whereas other forms of arts training did not.

Goldstein's first studies were published almost ten years ago, but since then, she and others have pushed her approach further. New studies have randomly assigned some participants to acting classes and others to "placebo" training sessions such as team-building workshops. Their findings, though preliminary, are promising. Medical residents who train in drama interact more empathically with patients afterward. Kids with autism who complete a two-week theater program perform better on empathy tests, and their families find interactions with them smoother.

As we've seen, empathy represents a push and pull between psychological forces. Acting shifts the balance of these forces. For one, it makes empathy more attractive. The ability to change perspective is theater's coin of the realm; actors who precisely model their characters' minds and hearts excel at their craft. Acting also reins in the costs that empathizing can impose, through the distance of fiction. Ella weeps while playing Belle, Orrie feels the wonder and confusion of a strange land as Alice. But their voyages into these fictional lives are short and voluntary. After two hours, they step outside into real life—perhaps relishing it more than they had before.

EVEN MORE LOW-KEY forms of untethering can build empathy. Playing Alice requires a crew, a stage, and a rare penchant for public speaking. Most people prefer to curl up on their couch and read about her instead. For more than a decade, the psychologist Raymond Mar has examined the effects of reading fiction. As he sees it, novels and stories give people a chance to experience countless lives. We can witness the struggles of a black woman in the Jim Crow South, or the isolation of pioneers on a lunar colony. We can strategize about what we'd do if suddenly granted the power to fly, or if we needed to break into Buckingham Palace.

Mar and others have found that avid readers have an easier time identifying others' emotions than people who read less. Young children who devour storybooks develop mind-reading skills earlier than their less bookish peers. Small "doses" of fiction encourage empathy as well. In one study, people who read George Saunders's *The Tenth of December* later identified others' feelings more accurately than people who had read nonfiction. In another, people were more likely to donate to an organization that supports depression research and treatment after reading a literary account of someone living with that illness, as compared to a scientific description of its toll.

Books are portable, silent, even secret; a reader can career through other worlds while their subway neighbor remains unaware. This means that readers can empathize safely even with outsiders they would disavow or avoid in public. We condemn murder, but we can spend hours in the mind of *American Psycho*'s Patrick Bateman. The son of a bigot can steal away with *Invisible Man;* the daughter of a homophobe with a copy of *Angels in America*.

These experiences are "contact lite": giving readers a taste of outsiders' lives, without the burden of real interaction. But they can still pave the way for caring about *real* outsiders. Its complicated legacy aside, Harriet Beecher Stowe's *Uncle Tom's Cabin* connected readers with the pain of a fictional slave, opening them up to the cruelty countless people actually endured. Its impact was so profound that when Abraham Lincoln met Stowe, he quipped, "So you're the little woman who wrote the book that started this great war." Upton Sinclair's *The Jungle* forced readers to confront the inhumane conditions of the meatpacking industry, sparking the movement for workers' rights.

This pattern pops up in laboratory research as well. In one study, people's attitudes toward LGBTQ and immigrant communities improved after they read stories with gay or foreign protagonists. In another, people read a fictional passage about an Arabic American woman who fended off racist attacks. Others read a synopsis of the same story—containing the same beats, but stripped of the evocative dialogue and inner monologue. Readers of the full story exhibited more empathy and less prejudice toward Muslims than those who read the blander synopsis.

Fiction is empathy's gateway drug. It helps us feel for others when real-world caring is too difficult, complicated, or painful. Because of this, it can restore bonds between people even when that seems impossible.

BATAMURIZA COULD BE a direct descendant of Shakespeare's Juliet: smitten with a passionate, warmhearted man from the wrong family. Her lover, Shema, is a Bumanzi. He lives in a village one hill over from Batamuriza's own community, the Muhumuro. Over the years, the Bumanzi receive preferential treatment from the government. The Muhumuro grow jealous. Batamuriza's brother, Rutaganira, agitates for violence against the Bumanzi. The villages descend into conflict, Rutaganira is jailed, Batamuriza joins a convent, and Shema attempts suicide. The Capulets and Montagues barely had it worse.

Musekeweya, or *New Dawn,* is a radio drama created in the wake of the Rwandan genocide. *New Dawn*'s creator, George Weiss, is the Belgian son of Holocaust survivors. He has spent his life trying to cure the kind of hatred his family endured. His tool is mass media, and he draws inspiration from people who used it to do harm. "The only model we have is what the bad guys did . . . the work of great propagandists like [the Nazi leader] Joseph Goebbels." Propagandists create fear and confusion, and then offer community and safety to those willing to toe the line.

This is what happened in Rwanda. In 1994, long-standing tension between the Hutu majority and Tutsi minority boiled over in the assassination of President Juvénal Habyarimana; the following day marked the beginning of an ethnic cleansing campaign that would last more than three months. By its end, 70 percent of Rwandan Tutsis had been murdered. On average, between twenty to forty people were killed every hour during the genocide.

The violence swept over the country "like the rain," as one survivor described it. But it had been building over time, in part spurred by savvy propagandists. The historian Jean Chrétien describes the genocide as driven by two tools, "one very modern, the other less so . . . the radio and the machete." Radio is by far Rwanda's most popular medium; across the country, people gather by the dozen to listen to music, headlines, and soap operas. In 1993, a new station, Radio Télévision Libre des Mille Collines (RTLM), took over the airwaves. While the national channel aired classical music and newsreels, RTLM played dance hits and gossip shows. It hired charismatic announcers, whose monologues made up more than half of the station's content. They joked constantly and shouted out people and villages they had visited. "Our priority is to assist all Rwandans by giving them their favorite news, laughter."

Many Rwandans remember these programs fondly, but RTLM had sinister aims. Its radio personalities were the PR arm of the growing, hateful Hutu Power movement. They falsely accused Tutsis of corruption and violence, and dehumanized them as *inyenzi*, or "cockroaches." After the violence began, RTLM took on a vicious tone, directing Hutus to "do the work" (code for killing), and celebrating their success. "Brew some beer for us to enjoy ourselves," an announcement rang out, one week into the genocide, "because we are seriously winning the war that . . . *inyenzi* and others have launched against us."

RTLM had opened a side door to hatred. Could radio also bring people together? This was the question *New Dawn* posed when it aired, a decade after the genocide. Around the same time, Rwandan officials, overwhelmed by prosecuting thousands of *génocidaires,* turned to a traditional justice system known as *gacaca* courts. Loosely translated as "justice among the grass," *gacaca* allowed victims to face their attackers in community tribunals; suspects could confess and apologize, and receive a standard sentence proportional to the severity of their crimes. But *gacaca* also publicly reopened psychological wounds, as people were forced to relive their trauma.

Weiss took a different tack. In the wake of the violence, he reasoned, people were too fragile for conversation. "We didn't want to talk about Hutus and Tutsis." He designed *New Dawn* to give Rwandans a safe entrée to think about betrayal, violence, and forgiveness. It also made room for empathy. The story's villain, Rutaganira, is a "transitional character." In jail, he finds new purpose, becoming a peacenik rather than a warmonger. Rutaganira's transformation highlights the idea that even killers are still human, and still have the chance for redemption. "The story shows that anybody can be a perpetrator," Weiss says. " 'They're not monsters. . . .' Or rather, since our audience is 90 percent Hutus, 'We are not monsters.' " Even if they weren't ready to forgive their neighbors—or themselves—*New Dawn* listeners could practice what it might feel like to do so.

At least that was the idea. We might not know if *New Dawn* had actually worked if not for an enterprising young psychologist. Betsy Levy Paluck—then a graduate student at Yale—had studied the poisonous effects of propaganda and was inspired by the idea that it could also be a force for good. Upon learning about *New Dawn* from Weiss, she impulsively offered to evaluate its effects, blissfully unaware of the years of effort that would entail. "Ignorance in these kinds of situations promotes bravery," she recalls. Weiss agreed, and Paluck started planning one of the most creative psychological experiments of the past few decades.

Before *New Dawn* was broadcast nationwide, Paluck organized "listening parties" in villages, survivor communities, and *génocidaire* prisons across the country. She played *New Dawn* for some groups, and a different radio soap opera—focused on health—for others. It was like a clinical trial with a story substituted for a medication (and more fun than taking a pill). Paluck ensured that her listening parties resembled everyday Rwandan radio experiences: a crowd gathered, drinks in hand. After the show ended, listeners danced, argued about the characters, and proclaimed support for their actions. "It is the aftermath of the program that is the most important," Paluck wrote one night in her field notes.

Paluck found that *New Dawn*, compared to the other program, increased listeners' empathy for people on both sides of Rwanda's tragedy. "People related to the radio story in an emotional and generous way," Paluck remembers. Even *thinking* about *New Dawn*'s characters could inspire empathy for real people. In a clever experiment, psychologists recorded the actress who played Batamuriza discussing reconciliation. Rwandans who heard Batamuriza's voice, as opposed to a voice they'd never heard, expressed more trust toward Rwandans of other ethnicities.

In Paluck's study, *New Dawn* did not erase the past. Its listeners weren't any more likely to support marriage between Hutus and Tutsis, for instance. Remarkably, though, the radio drama did increase people's sense that *others* supported reconciliation. Media often moves the needle in this way: first changing people's impressions of their community's beliefs and only later working its way into their own. To Weiss, this is especially true of Rwanda. "It is a very group-based society. People don't talk as much about their personal attitudes, but look to those around them."

Paluck's work suggested that a "dose" of *New Dawn* could soften Rwandans' fear and anger. Soon its medicine was spread across the entire nation. To Weiss's surprise, *New Dawn* became the most popular radio drama in Rwanda's history. At one point, 90 percent of the country tuned in to each episode. Over the years it has provided many moments of collective healing. Several seasons into the show, Batamuriza and Shema marry, finally joining families from the Bumanzi and Muhumuro. The wedding, though fictional, was recorded live at Amahoro Stadium in Kigali. More than a decade before, this venue had housed tens of thousands of Tutsi refugees. Now Tutsis and Hutus gathered there to celebrate.

New Dawn also helped people through difficult moments. During *gacaca* trials, victims often filtered their experiences through *New Dawn*'s lens. "People would talk about real people and real situations, and call them by the names of characters," Paluck recalls, "saying 'she was such a Batamuriza,' meaning she was wanting peace, or 'he was a Rutaganira,' meaning that he was inciting violence. It gave people a language to talk about the violence . . . and people's roles in that violence, but without having to directly accuse. The story gave them a way out in a pressure cooker of a moment."

Paluck never thought that a soap opera could erase the trauma of genocide, but she did believe that it could help Rwandans start to heal. "I can't say that this radio program led to true forgiveness or true reconciliation, but I hope it allowed their minds to wander in that direction."

LIFE AFTER HATE surprised me because it didn't focus exclusively on overturning members' prejudice toward outsiders. It started by building their compassion for themselves. Likewise, novels, plays, and fiction can help readers recast their own lives through the characters they meet, especially when they desperately need a new story to tell.

Changing Lives Through Literature was conceived in 1990 over a tennis match between two angry Bobs. Bob Waxler, an English professor at the University of Massachusetts–Dartmouth, saw literature becoming marginalized—as a creaky luxury compared to engineering and computer science. But each semester, he also saw students draw real-life meaning in the pages of classic novels. "I started thinking, 'I'd like to find some way to demonstrate that literature still has the power to make a difference.'"

Bob Kane, a judge at New Bedford, Massachusetts's District Court, was also frustrated. The same people arrived at his bench over and over again, charged with the same crimes. Kane's court felt like a revolving door. The Bureau of Justice Statistics recently tracked more than four hundred thousand prisoners released in 2005; by 2008, about two-thirds of them had been re-arrested. Once people are in the justice system, the data suggest, they have trouble getting out.

The two longtime friends vented together, and then Waxler proposed an unusual experiment. Kane would select convicts and shorten their jail sentences, but only if they agreed to join a reading group led by Waxler. Kane agreed immediately ("I loved it") and recruited Wayne Saint Pierre, a probation officer, to find candidates.

The three agreed on a few requirements. Prospective students should have long rap sheets and a high risk of reoffending—"Make them tough guys," Waxler said. Saint Pierre quizzed them on a paragraph from *National Geographic* to make sure they could read. When Kane offered places in the program, some men hesitated. "They knew jail," Waxler recalls, but many had never read a novel or been on a university campus. But ultimately almost everyone agreed, and the first Changing Lives group was launched.

The class would meet every two weeks in a UMass–Dartmouth seminar room, and spend an evening discussing novels such as *The Old Man and the Sea* and *Bastard Out of Carolina*—stories of risk, loss, and redemption. After the first session, Saint Pierre and Kane would join the group. Convicts would discuss literature alongside the judge who had sentenced them and the probation officer who could send them back to jail. Students would be responsible for finding assigned books at the library, and failures to show up to class or do the reading would count as parole violations. This was not your average book club.

Waxler arrived at Changing Lives' first session excited but apprehensive. After being announced, the program received some interest and a truckload of criticism. The UMass administration balked. "The first thing I heard from them was, 'You're bringing criminals onto the campus, and they're going to steal all our computers.'" State officials complained that Changing Lives was offering free education to convicts instead of people who deserved it. To many, the entire idea seemed like coddling, not justice. If just one student reoffended in a high-profile way, the program would probably be scrapped.

Waxler got the tough guys he asked for. His first class was made up of eight students, with 142 convictions among them, including several for violent crimes. Waxler convened the group: "I looked at one guy and said, 'You look a little tense,' and the guy looked at me and said, '*You're* the guy who looks tense, Professor!'"

Waxler handed students printouts of "Greasy Lake," T. C. Boyle's short story about three teenagers and a night gone awry. Buzzed and driving a Chevy Bel Air belonging to one of their moms, the protagonists arrive at a clearing near an old swimming hole now covered in a film of oil. They mistakenly interrupt a couple mid-romance, leading to a brutal fistfight. Things escalate quickly: The three teens potentially (but not clearly) attempt sexual assault and end up hiding out in the fetid pond while their car is trashed in retaliation.

After half an hour of silent reading, the discussion began. It was awkward at first. "They didn't have a good sense for the rhythm of a literary conversation," Waxler remembers. Some went long and others said almost nothing. Waxler, trying to steer the group, hit on a question about the characters. "Are these bad guys? Or could this have happened to anybody?" This launched a discussion about moral ambiguity. These characters had acted badly to be sure; had things gone another way, they could have done something much worse. But the story wasn't about villains. It was about how mistakes build on themselves, and how quickly people can lose control.

As the class streamed out, one student told Waxler, "This story is really my story." That's when he knew the program was onto something. Changing Lives discouraged students from talking about their own biographies—but the characters in each story gave them a new lens through which to see themselves. Many of the students had been called "bad guys" for most of their life, and had scarcely been given a chance to be anything else. Fiction revealed that underneath every crime is a person: flawed but still deserving dignity.

Over the weeks—and then years—Waxler saw fiction open up his students in other ways. It gave them a sense of possibility. Though a character might lose everything, the students themselves still had a chance: "The tragic hero comes to some realization about himself which he regrets. It's too late for him, but it's not too late for the reader." Through stories, students could imagine alternate futures, and paths to get there. "It allows these guys, who are often stuck in the perpetual now, to break out of the present moment and reflect back on their past, and what they might be able to do to create a future for themselves."

Changing Lives' students were often surprised at how differently they interpreted identical material. Through discussion, they learned that the same character could come off as sinister or naïve, depending on the reader's perspective. Waxler instilled a democracy around the seminar table. No matter what they'd done or how they were treated outside, here their opinions mattered.

This became even clearer when Judge Kane, who had sentenced these students, joined the group. To many of them, he embodied the system that had written them off: the man in a long black robe looking down at them from his bench. Some students didn't want him there at all. But Kane was not there to judge anyone. He read the books and shared his interpretations. More important, he listened to the students, responding to their feelings about a character or asking follow-up questions. Here they weren't pleading their cases; they were trying to figure out the motives of a Toni Morrison character. The sheer experience of speaking as equals with a judge—about pain, loss, and love—was a revelation.

At the end of the course, Waxler and Kane held a graduation. Students returned to Kane's courtroom, many joined by family and friends, to receive diplomas and books. The celebration recast them as people with insight and opportunity, in the same room where they had been condemned just months before.

The course ran a second time, then a third. Students started telling stories, not just reading them. One—who had neglected his family while dealing drugs—began reading to his three-year-old daughter. He told Waxler, "It may be too late for me, but not for her." Another was feeling weak one night after years of being sober. He walked down Union Street, New Bedford's main drag, peering down the side street where his old dealer lived. He found himself thinking about Hemingway's *The Old Man and the Sea*. In it, a fisherman named Santiago comes home empty-handed every day for three months, but perseveres. Waxler's student later told him,

"If he could do what he did . . . then I can walk straight down Union. I don't have to make the turn."

"He might not always listen to Santiago," Waxler reflects, "but Santiago will always be there for him, as a friend he can count on. . . . This is the power of what literature can do." Waxler and Kane believed, and other people began to believe, too. Judge-professor pairs opened new branches of Changing Lives in Lynn, Dorchester, and Roxbury—some of Massachusetts's poorest communities—and then in California, New York, and England. The program might be a liberal's dream, but it even took root in Brazoria, Texas, a county that overwhelmingly favored President Trump in the 2016 election. Texas is law-and-order country; between 1976 and 2015, more prisoners were executed there per capita than any other state except Oklahoma. Yet judges there saw the upside of giving criminals a second chance. The alternative was grim. "If Changing Lives doesn't work," one quipped to Waxler, "we'll kill them."

Researchers pulled the records of Changing Lives' first four classes—thirty-two students who entered the program with an average of eighteen convictions each—and matched them to a group of men on probation who were similar in age, race, and criminal history. By the end of the year, 45 percent of probationers in the comparison group had reoffended; five of them had committed violent crimes. In that same time, less than 20 percent of Changing Lives students had reoffended, and only one had committed a violent crime. Even students who reoffended often committed less serious crimes than they had before. Waxler thinks this is because Changing Lives instilled greater empathy in them. "I like to believe that . . . they have a little more respect for human beings, and have to think twice before hitting another person." A more ambitious follow-up tracked about six hundred Changing Lives students, along with a similar number of comparison probationers. Again, students were less likely to reoffend, and when they did, they committed less severe crimes.

Changing Lives is the first program of its kind, and research on it is still preliminary. But for many people, it's working, and it's hugely cost-effective. Changing Lives costs about $500 per student, compared to upward of $30,000 for a year of jail time if that student reoffends. And that might be its least important metric. When the program launched, Judge Kane was accosted by a victims' rights group. How could he put criminals back out on the streets before they'd served their time? He replied, "Imagine how many new victims there *won't* be because of this program."

Since Changing Lives began, more judges have integrated literature into criminal justice. In 2008, twenty-eight young men and women were convicted of trespassing and vandalism after holding a raucous party in Robert Frost's summer home in Vermont. Instead of jail time, they were sentenced to a custom-designed seminar on Frost's life and work. In Brazil and Italy, prisoners can shave three to four days off of their sentence for each book they read.

In many ways it's clear that we, as a society, still view literature as superfluous. In 2006's *Beard vs. Banks,* the U.S. Supreme Court upheld the right to deny prisoners reading materials, despite evidence that prison libraries increase inmates' likelihood of finding work once they're released. To Waxler, decisions like this signal our culture's obsession with the bottom line, which places arts at the margin. "Based on the growing notion that the human being is really an economic being, humanities and literature really have . . . very little room to do anything." It's hard to disagree with him. The National Endowment for the Arts has steadily lost funding and is at risk of being eliminated altogether. In 2014, 28 percent of all New York City schools lacked a full-time arts teacher; in poorer neighborhoods, that number was over 40 percent.

The arts have trouble demonstrating their value in terms that policy makers appreciate. A biochemist can quantify their contribution more easily than a playwright. But teamwork between psychologists and artists is changing that. When Thalia Goldstein launched her study of theater, she ran into skepticism from her fellow scientists but enthusiasm from actors and drama teachers.

"They wanted evidence. Evidence that what they're doing is good for people. Evidence that they should exist."

That evidence is now flooding in. Art—especially in narrative forms such as literature and drama—helps us untether. It makes empathy safer and more enjoyable, even in the hardest circumstances. Storytelling is among our oldest pastimes, and, as it turns out, one of our most essential.

CHAPTER 5

Caring Too Much

THE UNIVERSITY OF California's San Francisco Medical Center at Mission Bay cost about $1.5 billion to build. The heart of that campus, Benioff Children's Hospital, is among the most advanced in the world. Its external walls are mirrored, dotted with stained-glass accents. Ambient music plays in the hallways. The art collection rotates often. Walking toward it, I feel nauseous.

The day my daughter Alma was born was one of the best of my life. It was also the worst. After a long and difficult labor, Alma entered the world in a Benioff operating room just before 2 A.M. My wife and I listened for her cries, but the room was silent. The doctors' and nurses' faces were shot through with concern. Alma struggled to live. We later learned that during birth, she had suffered a stroke. She was rushed to Benioff's intensive care nursery (ICN), where she lay shell-shocked under a heat lamp.

In Alma's early moments I learned two things: First, I wanted to protect her more than I'd ever wanted anything. Second, I had already failed.

A parade of medical staff visited us at Alma's bedside. They came at 5 A.M., at noon, at midnight, sometimes for a minute, sometimes for twenty, usually without warning, always with news. Signs of infection had dissipated, but the inflammation under her skull had not. Her seizures were not life-threatening, but they might continue for years. The doctors merely translated Alma's charts, but it felt to us as though they had power over what was in them. After a punishing set of results, we'd wish for their mercy.

What the ICN staff did control, though, was how they treated us. They answered all our questions and sat with our worries. After delivering some bad news, one doctor spent a dark early morning hour talking with me about fatherhood. Our real lifeline was Liz Rogers, Alma's neonatologist and the unit's associate medical director. Liz's hair is a mosaic—chestnut, streaked with gray in some places and highlighted blond in others. Her face is a mosaic, too. While she talked about Alma, she smiled, but her eyes seemed sad, searching ours. She hugged my wife and me every time she entered Alma's room. She cried with us, and talked about her own children.

I've studied empathy for years, but I have rarely received it in such a profound way. These doctors, nurses, and technicians were strangers, yet they became the closest people to us at the hardest moment in our lives. They've done the same for countless others. The ICN specializes in treating vastly premature children, born at a knife's edge between life and death. Some are so delicate that raising their legs could cause their brain to bleed. Families here face a fear that most parents cannot imagine; if sadness were light, you could see the ICN from space.

Each day, Liz and the doctors, nurses, and staff witness and work against this misery. Then they go home to their own families and act as though everything is fine, and return the next day, ready to give again. They are like empathic superheroes. But can they possibly sustain this rhythm and, if so, for how long? What does their care cost?

THROUGHOUT THIS BOOK, we've seen people benefit from turning up their empathy. But imagine taking on everyone else's feelings, all the time. You couldn't walk one Manhattan block or watch the news without collapsing into a heap. If your son broke his leg, you'd panic so much you'd be useless. If your friend broke down, depressed about his divorce, you'd cry too much to comfort him. You'd make a lousy therapist.

No single emotional experience is always helpful or harmful. Anxiety feels bad, but it can energize us to face challenges. Joy feels good, but at its deep end it can turn into mania. We could all use more empathy sometimes, but too much of it can debilitate us. Twenty-five years ago, Carla Joinson coined the term "compassion fatigue"—empathy's repetitive strain injury. "Human need is infinite," she wrote. "Caregivers tend to feel 'I can always give a little more,' but sometimes they just can't help."

Joinson focused on nurses, but compassion fatigue can wreck anyone who's constantly surrounded by suffering. In the United States alone, that includes tens of millions of people. Those who suffer from chronic illnesses receive most of their daily care from a relative. Caring for a spouse or a parent is a moral responsibility that can be rewarding; it's also stressful. Caregivers watch as their loved ones suffer pain and disability, helpless to stop it or predict what will happen next. In many cases, they must be constantly on guard for crises, and as a result have trouble working and maintaining relationships. This wears on them. Caregivers are much more likely than others to develop depression, and they also report worse overall health.

This is true for parents, too. We tend to see exhaustion as a defining feature of good parenting: the belief that one has given their child everything and then a little more. But living up to this ideal can take a physical toll. In one set of studies, parents reported on their empathy, and their adolescent children described their own emotional lives. Kids of empathic parents reported less anger and fewer mood swings, and bounced back more quickly from stress. But empathic parents paid a price: They exhibited low-grade inflammation and greater signs of cellular aging. When adolescents experienced depression, empathic parents (but not less empathic ones) suffered even more inflammation. It was as though caring parents transferred their own well-being to their kids.

Even without children or ailing loved ones, people can easily become stretched by caring too much. Modern life gives us unprecedented opportunities to broaden our empathy, and journalists and activists have taken full advantage, inundating us with images of suffering in the hopes of inspiring action. This logic is behind ads from Save the Children and the ASPCA: Seeing a hungry child or a defenseless animal, how could anyone *not* care? How could anyone not help?

But by inundating us with these images, the media fosters an epidemic of compassion fatigue. In 1996, psychologists found that up to 40 percent of television viewers were exhausted by coverage of bad news. Since then, the news cycle has quickened and intensified. In minutes, an addled reader can click through stories of mass shootings, children separated from their parents at the U.S.–Mexico border, and natural disasters ripping through the Caribbean. In 2018, a Pew poll found that almost seven in ten Americans experience "news fatigue."

Few are at greater risk of overdosing on empathy than "caring professionals": physicians, social workers, therapists, teachers, and others who work with people in need. As we've seen, when people feel like someone else's pain will overwhelm them, they steer clear. Caring professionals don't have that option. And even if they did, many wouldn't want to. They are humanity's first responders, called to their work by a deep concern for others. To run from pain would betray their core values. But in empathy's trenches, those values can quickly turn into occupational hazards.

EIGHTEEN MONTHS AFTER Alma was born, I return to the ICN to observe the workers there. At the beginning of the day, about half a dozen nurses, medical students, and residents rotate through the unit, discussing each patient. These rounds, helmed by an attending physician, last for two to three hours. Everyone but me clearly chose their shoes for comfort: an even split between sneakers and padded clogs.

Melissa Liebowitz is a fellow at the ICN and will soon become the unit's newest attending physician. Today she is leading rounds, shadowed by Liz. Melissa has been in this role all of two weeks, but it might as well have been two years. She exudes wisdom and patience. As medical students and nurses run through updates, she peppers them with questions. She also looks tired. Her eyes water, and she shifts her weight from side to side.

As they discuss each case, the team uses rapid-fire shorthand. The first baby we see was born at twenty-five weeks, weighs seven hundred grams, up fifty from yesterday. Last night he had twelve As, four Bs, and three Ds—episodes of apnea, bradycardia, and desaturation: common malfunctions of premature hearts and lungs.

These details describe the baby in front of us, but they obscure how much he's struggling. Seven hundred grams is about a pound and a half. He wriggles and cries at an impossibly high pitch. A nurse reaches into his Isolette—or incubator—through a plastic barrier. His arm is roughly the size of her middle finger. He is surrounded by intricate machinery, hooked to him in a dozen places. He looks like the cell of a battery, like he's powering the devices instead of the other way around. Given his gestational age, he has about a 70 percent chance of surviving the year. A dozen babies like him surround us in this wing of the ICN; the entire unit houses about fifty-five.

Standing with Melissa and her team, it's hard for me to remember what this place once meant to my family. The pastel-colored flower murals and uncomfortable vinyl chairs ring bells. Some staff members seem familiar, but only vaguely, like characters from a dream. My phone remembers the unit's Wi-Fi. I wait for the dread I felt outside to build, but instead it breaks on a wall of acronyms. For the staff here, too, the constant details can keep feeling at bay. "It's a protective mechanism," Liz tells me, "and maybe also a way to avoid." Alarms go off constantly, with varied, soothing tones. They startle no one.

You can guess a baby's prognosis by how long the team stays at his or her side. If we pass by quickly, chances are the kid is stable. If the staff lingers, something's wrong. Today, we spend the most time discussing Francisco. He was born three weeks ago in a small hospital near San Jose, twelve weeks premature. After his first week or so, Francisco stopped eating, his belly swelled, and his stool turned bloody. These are hallmarks of necrotizing enterocolitis, or NEC: a poorly understood illness in which a premature child's gut dies inside their body. The team will operate later today to examine the extent of the damage.

Francisco's case makes the team visibly nervous. Last night he had a string of Bs, and they needed to jump-start his heart using CPR. They never say so, but as they talk, it dawns on me that he is not likely to survive. A baby dies at the ICN about once a week, but still it seems impossible that *this* baby—or any one in particular—would be the one to go. The group disperses, and Melissa stays to examine him. His skin is the color of a walnut shell, his abdomen distended and translucent, blood vessels visible underneath. His black hair is matted and wet. His eyes are closed, but his face, through the tubing, is contorted with effort. His right hand is balled into a fist the size of a blackberry. "This is the type of exam that hurts my heart in a new way now that I have a son of my own," Melissa tells me.

Melissa's son is five months old; her work has changed the way she experiences motherhood. She worries more, and more specifically. Once, after her son spit up several times, she examined him for pyloric stenosis—an intestinal condition—until her husband groggily pleaded for her to stop. Melissa waited longer than most expecting mothers to tell her parents and friends about her pregnancy, intimately aware of all that could go wrong.

Before Francisco's surgery, we receive more bad news. An

ultrasound revealed a hemorrhage inside his brain, worsening his prognosis. The surgeon wonders aloud whether it's still worth operating. "Now we've got a kid with a head bleed and no gut," she announces. "We need to have a powwow." Elana Curry is the ICN social worker handling Francisco's case; Liz and Melissa speed-walk to her office, because things are about to get complicated. Francisco's parents are migrant field workers. They speak only a Mexican indigenous language and have no experience with complex medical issues. Elana explains to them, through a translator, how antibiotics work, what's involved in surgery, and what intestines are.

The team discusses whether to operate on Francisco. "If it were my baby, I wouldn't do it," Melissa says. Even if this surgery succeeds, Francisco will likely be machine dependent for the rest of his life. Melissa sees nothing but more suffering ahead for him. The team wants Francisco's parents to decide, but how can they, given how little they understand? Elana scrambles to think of ways to describe things like hemorrhage, prognosis, coma. The team holds a private "family meeting" with Francisco's parents. I wait outside the conference room. Inside, a rectangular wooden table is surrounded by rolling chairs, paper cups stacked upside down on top of a cabinet beside it. My wife and I cried bitterly at that table a year and a half ago.

They emerge after just a few minutes. "That was one of the harder conferences I've had in a while," Liz reports. Francisco's parents didn't grasp what was happening and wouldn't make a decision. They ceded the choice to Liz, who decided to move forward with the surgery. One of her imperatives is to empower families in their most helpless moments; making the choice for them feels all wrong.

As soon as the surgery is confirmed, a dozen people materialize, full of purpose. They measure, adjust, and operate. The surgery occurs in Francisco's room in near silence. It's shocking how unaffected the rest of the unit is: a dozen feet away, two parents await their son's MRI results, wearing sweats from the night before. In the next room, a janitor mops underneath an Isolette while a nurse peeks into it. Behind them, a mother lies curled up on a love seat. In the main hall nurses and medical students decide which type of curry to order for lunch. It seems cruel for them to act as though things are normal—but here, this is normal.

I go to the cafeteria and try to eat but manage only to get into a staring contest with a paper tray of chicken fingers. Less than five minutes later, Liz texts. "Jamil, sounds like they are finishing without good news." I realize I haven't seen Liz eat, use the bathroom, or take any other break in the past five hours. Upstairs, Melissa emerges. During surgery the team discovered that Francisco's entire intestine had died. His condition is downgraded from NEC to "NEC totalis," which is described as "incompatible with life." Melissa is most upset by something else; when they made the first incision, Francisco's heart rate spiked, an indication that he'd felt pain.

THE HOSPITAL'S ROOF terrace is dotted with pale, wispy shrubs. Wind cascades over a nearby hill and rolls over us before flowing out into the bay. We have an unobstructed view east: scattered piers, barges drifting imperceptibly to or from Asia, Oakland. Francisco's parents are tiny in this great expanse. They sit on a chrome bench holding each other, faces turned down. Seven ICN staff members emerge. They yank at other benches on the terrace but find them bolted down, and decide to kneel instead, encircling the parents like worshippers at a shrine.

The surgeon speaks through an interpreter, both of them conveying the news in a near whisper. Melissa tears up, managing only to get out "I'm so sorry." Liz is more eloquent. "Although we can no longer decide whether he lives or dies, we can choose how he spends the rest of his time, and we want that to be with you." Francisco's mother drops her head into her hands. Her husband holds her, nodding forcefully at everything the translator says.

The team stays for as long as Francisco's parents want, but that's not very long. His father asks if his son's gut is really dead. It is. But his heart is still beating? It is, but only because of the machines. Liz asks if they should wait for more family to arrive before removing his life support. Everyone is working, the parents report, and no one can come. "We'll leave you to think," Liz says. Each staff member places their hands on the parents before retreating. Outside, Francisco's mother and father sit quietly in the wind. The ICN team reenters the hospital, and their swarm breaks apart.

ICN WORKERS LIVE in a prism. The same facts go into each room: A baby is born, but things have gone poorly. What comes out is a set of alternate realities, each one totally enveloping the family living it and inaccessible to every other family. I can no more fathom what Francisco's father is going through than a father who has never seen the inside of a neonatal intensive care unit can understand what I went through. ICN parents do share some things, like exhaustion and dread. As Samantha, an ICN nurse, puts it, "It's the end of the world for all of them."

When Alma was a patient here, I saw these doctors and nurses as all-powerful. In truth, they are as vulnerable as the rest of us. In a few steps, they can move from a dying child to one who will thrive to one whose future is unknown. They cross vast emotional terrain in seconds, pulled along by others' need.

In a space like this, empathy carries many risks. One psychologist who has studied medical empathy for decades writes that "understanding binds, but feeling blinds." He counsels caring professionals to adopt what he calls "detached concern"—a Vulcan sort of goodwill. More than a coping mechanism, this is the industry standard, according to the oncologist Anthony Back: "In the idealized professional model," he writes, "physicians' feelings are extraneous." Liz rejects this model. "That to me would not be a satisfying way to practice. . . . Being honest and open about how a patient has affected you as an individual can only give more meaning to them." She's seen dozens of infants die, and she cries every time.

But emotional investment can bring caregivers to strange places. Samantha once spent seven months treating a baby who had been abandoned at birth. She began buying outfits for him and thinking about him constantly even when not at work. When he became healthy enough to join a foster family, Samantha was happy for him, but also heartbroken. What's worse, she found it hard to be present for her other patients. "I kind of resented them for not being the baby that I wanted to be taking care of." She was ashamed of feeling that way ("It's the opposite of what you're there to do"), but for a time, her connection to one baby had overwhelmed everything else.

Over-empathizing can tempt caregivers to take heroic lifesaving measures even when they are likely to fail, or to sugarcoat bad news to avoid causing pain. One ICN doctor tells me about a baby he is treating who will likely die. The multispecialty team has hinted at this, but not clearly enough for the parents to truly get the prognosis. "They're such nice people," the physician says, "and you don't want to tell them such bad news."

Empathic caregivers can also develop PTSD-like symptoms— not from their own pain, but from their patients'. A quarter of nurses in neonatal intensive care report "secondary trauma"— sleeplessness, flashbacks, and exhaustion—at about twice the rate of nurses in other specialties. Secondary trauma often gives way to burnout: general exhaustion and loss of meaning. One in three intensive care professionals suffers burnout, a much greater proportion than in other branches of medicine. Empathic professionals bear the brunt of these problems. They become depressed more often than their less empathic peers, and they are more likely to blame themselves when patients worsen or die.

I see burnout, fatigue, and trauma all over the ICN. After a string of losses at the unit, Melissa developed symptoms of depression and anxiety. "I was coming home most nights," she recalls, "sitting on the couch and crying a lot." She now feels stronger but still can't watch the news. "With the thought of other suffering in the world, on top of this job, how could I ever enjoy life?" I ask one ICN nurse how he deals with his emotions. "I just push them down until it becomes a health problem," he replies, laughing. Months later, I run into another ICN nurse at a bar and ask her the same question. She points to her drink.

At the outset of their training, medical and nursing students score higher on empathy tests than people starting other careers. In many ways, that's a good thing. Patients of empathic physicians tend to be more satisfied with their care, are more likely to adhere to medical recommendations, and even recover from illness more quickly than those whose doctors are more detached. But these same qualities can compromise caregivers' own health. This puts empathic professionals in a double bind. Some continue pouring themselves into their work, but they risk ending up with nothing left to give: burning out, quitting, or both. Others simply shut themselves off. In the first weeks of their training, medical students are preternaturally empathic; by their third year, they empathize *less* than the general population. This affects the care they give. Physicians and nurses underestimate patients' suffering, and tend to dehumanize patients—viewing them less as people and more as bodies.

Disconnection can help caregivers survive their work. In one study, doctors and nurses were told the story of a terminally ill patient, and asked to guess how that person might feel. Individuals who dehumanized patients experienced less burnout in their work. But self-preservation comes at a cost. If patients benefit from empathic physicians, it follows that they'll suffer at the hands of detached ones.

There's a dangerous cycle of empathy, exhaustion, and callousness in medicine, and it's accelerating. Managed care—in which medical costs are closely monitored by insurers—has exploded, increasing workloads and squeezing caregivers' time. Primary care physicians speak with patients for less than fifteen minutes per visit, and typically interrupt them within the first thirty seconds of conversation. Between 2011 and 2015, rates of physician burn-out crept upward by 10 percent. And as marathon twenty-four-hour shifts become the norm, caregivers become more burned out, less empathic, and less willing to listen to patients and their families.

The ICN's staff successfully buck this trend, finding a way to make families like mine feel heard. But they bear a heavy load.

FRANCISCO'S PARENTS DECIDE they can't be there when his life sup-port is removed, but at the ICN every baby dies in someone's arms. Today, Elana volunteers. She sits on a love seat as a nurse begins freeing Francisco from the machines around him, beginning the end of his life. A dozen people encircle her and Francisco. Most are standing, but Liz sits next to them, and Melissa kneels at their side. I imagine we look like mourners in a renaissance painting, our bod-ies angled toward a focal point. Above us, there's a printed cartoon of a baby Mickey Mouse with the name "Francisco" written below it. Liz closes her eyes and places two fingers on Francisco's forehead. His ventilator is switched off, filling the room with silence.

The team checks and rechecks to ensure Francisco is receiving enough morphine. Liz and Melissa occasionally feel for his heart-beat, which continues for more than twenty minutes after his last breath. The fourth time she checks, Melissa looks up to the clock, tears streaming down her face. Everyone hugs. The team begins preparing Francisco's body.

Ten minutes later I'm three blocks away at a small park. Sun filters through the trees; a woman pushes a double stroller. A tod-dler starts a determined run before falling into the grass. My phone reminds me to pick up dry-cleaning, and every bit of it feels impossible.

Over the following days, Francisco returns to the ICN in

different ways. Melissa has trouble letting go of the fact his heart rate spiked during surgery. Did he feel pain? She can't know for sure, but she thinks she could have made a difference. "I should have been a better advocate for him. . . . The idea of him not knowing what was happening . . ." She trails off. Molly, another nurse in the room during his death, thinks about his parents. "They had nothing, and he was their something. And the fact that they didn't want to see him again or hold him . . . I get that, but that made it harder, too."

Each team member copes in their own way. Molly goes for long runs. "It gives you solitude, and helps you find peace with whatever you've seen." Melissa holds her son more tightly than usual at night. Liz sings along with the radio on her commute home, "Carpool Karaoke"–style. Sometimes she waits a few minutes in her car after arriving home, treating it like an oasis. "When I get home, I'm the mother of four kids who need me to be their mom. That 'in between' is the only time to transition from one world to the other."

These are all forms of "self-care," practices for resetting one's own emotions. Self-care certainly can't hurt, and in some cases it protects caregivers from burnout and fatigue. But in the face of immense pain, it might not be enough. In one study, therapists who used self-care suffered just as much secondary trauma as those who used it less. And even when self-care does make a difference, staff in the most stressful settings—such as NICUs and ERs—might not have the time or structural support to make use of it.

More important, counting on self-care to fix burnout diminishes how serious this condition is. When we're sick, we turn to professionals for a reason. No doctor would tell someone with internal bleeding to take a bubble bath, or someone with a broken wrist to watch a funny movie and let it pass. When we expect caregivers to get better on their own, we are in essence telling people who are in real distress to walk it off.

Caregivers do reliably benefit when they get help from others. Social support buffers nurses and doctors against burnout. The problem is that tragedy is isolating, and caregivers find it hard to reach out for help. I experienced that firsthand. In the forty-eight hours after Francisco's death, I saw two dozen colleagues, a few friends, and my own family. When they asked me how things were, my mind screamed, "Francisco died yesterday!" but I couldn't bring myself to say so. I'd chosen to be at the ICN, but none of my friends had. I couldn't impose this tragedy on them. On the other hand, their news—a paper had been rejected, a date had gone well—felt trivial. It wasn't my friends' fault, but Francisco had eclipsed my interest in regular life.

ICN staff members rarely disclose what happens on the unit, even to their loved ones. "People become hesitant to ask how work is," one nurse says, "and even when they do they don't know how to process it." The work here can also make it difficult to take other people's experiences seriously. The same nurse tells me, "Sometimes I come home to my fiancé and he's telling me about a tough marketing problem he had that day, and there's part of you that thinks, 'I don't care. That is *not* a problem.'" People who witness great suffering often see less of friends in other professions. If someone else can't understand what you've gone through, and you can't care about what they go through, what's the point?

The ICN staff support one another, but almost all of this occurs in the margins, during short breaks or over drinks after work. The team does hold debriefing meetings each time a baby dies. At Francisco's debrief, Liz provided fruit and yogurt for breakfast ("We eat too many doughnuts around here"). "We were able to sit and talk," she says, "and share whatever people wanted to reflect on." I ask Melissa if debrief is a good place to discuss her feelings. "Not so much," she responds. "Most of it is just the business. . . . We acknowledge people for good things they did, and share sadness, but don't really talk about how it's affecting us in the rest of our lives."

The ICN offers professional support, but the staff almost never use it. During one of her lowest points, Melissa was asked by a senior physician if she wanted to see a social worker. "Come on, I don't need that right now," she answered, as if the mere suggestion called her emotional stamina into question. She sees medical residency as a test of her mettle, "kind of like the military . . . people are expected to just get on with their lives."

RECENTLY, THOUGH, THIS has begun to change. Caregivers are becoming more purposeful about empathy—finding ways to combat burnout and better support one another. At Johns Hopkins in Baltimore, change began with a tragedy.

By eighteen months old, Josie King had earned the nickname "Wrecking Ball." She joyously dumped out boxes, scattered closets' worth of clothes, and danced as much as she walked. In January 2001, she snuck away from her parents and tried to draw herself a bath, scalding herself severely. She was rushed to Hopkins and treated at their intensive care unit. Her condition improved steadily; her siblings blew up balloons and made colorful cards to welcome her home. Then, with little warning, Josie suffered cardiac arrest. Within twenty-four hours she was gone.

Her death left a crater in her family, and as time went on, their grief mixed with rage. In Josie's last days, the Hopkins staff had made a series of errors that put Josie at risk of infection and dehydration—the eventual causes of her death. The hospital settled with her family, but Josie's mother, Sorrel, still fantasized about revenge. "They must suffer," she wrote in her journal. "They must feel the pain that we feel."

She also thought about what she owed Josie. "I will do something great for you," she wrote. "Please help me find out what that is." Sorrel and her husband, Tony, wanted Josie's death to help other children. In an astounding move, they donated part of their settlement to Hopkins to create the Josie King Patient Safety Program. Over the years, Hopkins overhauled their practices to prevent errors. Many of their new standards have caught on nationwide and have saved countless lives.

Around the same time, Albert Wu, a professor of health policy and management at Hopkins, had his mind on another side of medical errors. Cases like Josie's devastate families, but what about caregivers? Wu interviewed residents about their past mistakes and found that many of them suffered symptoms of PTSD. He realized that his hospital was full of victims. But unlike patients and their families, caregivers didn't allow themselves to ask for help. Their actions had caused pain; it would be self-indulgent to voice their own.

Wu saw a dangerous combination of vulnerability and silence surrounding errors. "Some of our most reflective and sensitive colleagues," he wrote, were "most susceptible to injury from their own mistakes." The truth was even worse than he realized. In one study, medical residents completed surveys every three months for a year, describing medical mistakes they had made, and their quality of life and work. After residents made an error, their burnout skyrocketed, their risk of becoming depressed more than tripled, and their empathy for patients decreased.

In 2011, Wu delivered a lecture on medical mistakes at Hopkins. When he brought up Josie King, two women walked out. He later learned they had treated Josie. Ten years later, the hospital had never asked them about their experiences. "It was a clear miss," and one that could be remedied. Wu and his colleagues felt that social support between caregivers should not be confined to hallways and happy hours—it should be an integral part of work. They created Resilience in Stressful Events (RISE), a hospital-wide peer-to-peer empathy network.

Wu's team identified "glue people" among the community, whom he describes as "wise, empathetic people who won't judge" their colleagues. The RISE team trained them in a counseling approach known as "psychological first aid," most commonly used after disasters. Victims and witnesses of earthquakes or terrorist attacks experience a flood of stress hormones. The world wobbles: In one moment it feels too focused; in the next, nothing feels real.

"Adverse events," such as errors or sudden deaths, affect caregivers in a similar way. As Wu explains, "They can tell you what the patient was wearing, what they [themselves] were wearing, what the weather was, the colors of the walls in the room."

These feelings get seared into "flashbulb" memories—a dangerous kind of untethering. When victims remember trauma, the fear and agitation they felt in the moment come rushing back. Psychological first aid attempts to short-circuit that process by providing victims with a sense of safety immediately after a disaster. Any Hopkins employee can call the RISE helpline at any time, and a counselor will respond within thirty minutes, usually within ten. Counselors listen and ask questions while withholding judgment, and occasionally point staff members to further help.

RISE opened its virtual doors in 2011, and almost no one came. In the first year they were lucky to receive one call a month. Physicians, especially, were slow to let themselves be vulnerable. But the stiff culture began to give way: Calls trickled in, from individuals who had made an error or lost a patient they thought would live. Soon medical teams were calling together. RISE now counsels upward of one hundred Hopkins staff members a week.

RISE offers caregivers a concentrated, bite-sized dose of empathy from their peers, but its effects reverberate. Recently, Wu and his team found that after an adverse event nurses who took advantage of RISE were far less likely to take days off or leave their job than nurses who did not. RISE takes the compassion that can exhaust caregivers and points it back at them, protecting their emotional lives. Programs like this can't solve every problem, and RISE has by no means eliminated burnout and fatigue at Hopkins. But it at least helps prevent caregivers from falling through the cracks.

Wu is in talks to spread RISE to all Maryland hospitals, as well as to sites in Texas, the Netherlands, and Japan. This is good news, but it also highlights how little social support many others receive. In more overworked settings, such as poorer hospitals or public schools, professionals likely have even less time for it. People who care in private—the spouse of an Alzheimer's patient, the father of a daughter with cerebral palsy, the best friend of someone in the throes of bipolar disorder—might not know where to find it. And even when they locate a community that can give them support, they need the courage to ask for it.

THE MISSION DISTRICT meditation center Against the Stream is a quintessentially San Francisco space. Rows of plush chairs surround a small wooden altar in an airy, cream-colored room. The walls are dotted with Shepard Fairey prints—mandalas surrounding his famous Andre the Giant "Obey" image: half religious iconography, half outsider art. Friday-evening meditation classes draw hundreds of people, but today there are fewer than a dozen of us here. The students—medical residents at Zuckerberg San Francisco General Hospital—sit silently, hands on their thighs, focusing on their breath. The room is peaceful, but we can hear a siren whiz by. For all we know, it's headed to the students' workplace.

Eve Ekman, who is leading the exercise, is a researcher at UCSF's Osher Center for Integrative Medicine and is pioneering a program to help doctors tune their empathy. She began this work by accident. A decade ago she was a social worker, handling night shifts at SF General's emergency room. This was depleting work, but Eve found solace in art and nature. "I saw levels of suffering that were hard to bear, and always made an effort to balance out the other side with beauty." She knew hospitals were using meditation to de-stress doctors, but she was unconvinced. "If you want to relax," she thought, "go grab a beer." In 2006, her father, the psychologist Paul Ekman, agreed to co-lead a teacher training program called "Cultivating Emotional Balance" alongside a Buddhist scholar. Paul fell ill just before it was scheduled to begin, and Eve agreed to pinch-hit for him, inadvertently changing her life.

We've seen the burden suffered by caregivers who go all in on empathy. But when they instead aim for detached concern, ignoring their own feelings, new risks arise. Caring professionals *do* feel, and "unexamined emotions," as Anthony Back calls them, pop up in harmful ways. Physicians who ignore their feelings make less accurate diagnoses and are more likely to take their frustration out on patients. Unexamined emotions also pollute the rest of caregivers' lives, making them more likely to lose sleep, clash with their families, or abuse alcohol.

The good news, according to Back, is that caregivers can work *with* their feelings instead of against them. He encourages them to turn their focus inward, diagnosing their own emotions the way they would a patient's illness. When a nurse meets a leukemia patient the same age as her daughter, she might experience an avalanche of sadness and should be on the lookout for signs that it's overwhelming her. To do this, caregivers must be attuned to precisely what they feel around patients. Psychologists call this ability "emotional granularity," and some people are better at it than others. In one study, people kept a diary for two weeks—each day, they reflected on the most intense emotional experience they'd gone through. How happy, amused, and joyful did it make them feel? How nervous, angry, or sad?

Some individuals revealed a sharp inner life: A fight with their partner made them very angry, slightly ashamed, and moderately sad. Others experienced their emotions in clumps. When a day was bad, it was just *bad;* every negative feeling pooled together. People who could pinpoint their feelings had an easier time controlling them and bouncing back from hardships. By understanding their emotions, they could change them, the same way someone with a detailed map can find their way out of the woods. High-granularity individuals benefit in other ways: They are less likely to binge drink, engage in violence, fall prey to depression, or harm themselves than people with a more opaque emotional life.

Scientists once viewed granularity the way they view empathy, as a trait people either have or don't. But new evidence suggests that people can learn to identify their feelings. One program taught schoolchildren a set of words that precisely described emotional states, and then helped them reflect on their feelings. Students who went through this program were rated as kinder and calmer by teachers, and their grades improved.

Paul Ekman spent his career mapping characteristics of anger, fear, surprise, and other emotions. His work made him a giant in psychology and also spilled into pop culture—the film *Inside Out* and the television show *Lie to Me* are both based on his ideas. Paul's work also meant that his daughter Eve grew up in a household that was highly aware of feelings. At the ER, she noticed differences in how her colleagues empathized. Some overdosed on it, growing cynical and exhausted. Others were nearly unflappable. "They had found this kind of balance between detachment and distress." Eve wondered: Can the rest of us become more like them?

Within days of stepping in for her father at the emotion and mindfulness training, Eve had her answer. The practices she saw went far beyond relaxation: They were an ancient technology for tuning emotional life. Meditation, she realized, "invite[s] people to be more curious about their experiences," and gives them a precise vocabulary for describing them. This included language for separating different kinds of empathy. In Buddhism, "compassion" entails caring about someone else without taking on their pain. "That separation," she says, "needs to happen, because if it goes too far one way it's 'that person, not my problem,' and if it doesn't happen, we can over-identify with the suffering around us."

Psychologists make a similar distinction, between empathic distress and empathic concern. Distress is one flavor of emotional empathy: feeling *as* someone else does by vicariously taking on their pain. Concern instead entails feeling *for* someone and wanting to improve their well-being. Concern and distress can seem like two sides of the same coin—if someone else's pain hurts you, you have every reason to help them feel better—but they need not travel together. They are only weakly related: Someone who experiences deep distress does not necessarily feel deep concern as well, and vice versa.

These states also inspire different actions. Easily distressed people avoid others' suffering, for instance, refusing volunteer opportunities that will put them in emotional situations. People who tend to feel concern do not. In one study, college students read about a tragic accident supposedly suffered by one of their peers, and rated how much distress and concern they felt in response. They then had the chance to help the victim. The researchers designed the experiment so that some students could easily get out of helping (they would never have to face the victim), whereas others could not. People who experienced distress helped when it was their only option, but stayed away when they could. People who experienced concern helped in either case.

In caring professions, knowing the difference between these states is vital. Distress motivates people to escape others' suffering, but caregivers can't do that without abandoning their post. This leaves them with a punishing psychological burden. In fact, of the different kinds of empathy, only distress tracks burnout among doctors, nurses, and social workers. Concern, on the other hand, gives them a way to emotionally connect with patients without taking on their pain, and caregivers who tend toward concern rather than distress are *less* likely to suffer from empathic injuries. In other words, empathy doesn't have to produce burnout at all, and experiencing the right kind might actually prevent it.

One day, I ask Liz if she's ever doubted her ability to go on in the face of the all the suffering she witnesses at the ICN. "No," she answers. "That's why I do what I do. . . . I enjoy taking care of people in crisis." She doesn't mean the enjoyment you'd get from an ice cream sundae, of course, but something deeper: the ability to help families through their toughest moments. A deeply spiritual person, Liz views those moments as "graceful, and beautiful."

"God created those situations," she tells me, "for a purpose and with meaning." Having spent time with her, it seems clear to me that she has a high empathic set point—geared toward concern—and this has probably helped her help families for years while holding on to her own optimism.

Gifted people often struggle to imagine being less gifted. A mathematician might not understand what it's like to not understand complex equations. I wonder whether Liz knows how hard this work would be without her stubborn hope. "Only recently have I realized that this is a particular trait not everyone has," she tells me.

If we accept the Roddenberry hypothesis, people with her empathic trait—or, rather, set of traits—are built for caring professions, and people unlike her should find other lines of work. But just as we can tune empathy up or down overall, we can tune ourselves toward different *kinds* of empathy. People without Liz's lucky set point might need help navigating intense situations without burning out, but they can also help themselves by changing their approach to caregiving.

As we saw earlier, the neuroscientist Tania Singer and her team have built people's empathy using Buddhist-inspired practices. They have also used this training to push people toward specific kinds of empathy. *Metta,* or loving-kindness meditation, cultivates empathic concern. In a recent study, Singer and her team trained one group of people in *metta* and another group in a meditation practice that focused on vicariously "catching" other people's emotions. After several days, people who practiced *metta* became more generous and less distressed than those in the other group. These changes showed up in their brains as well. People who had learned to share others' pain exhibited heightened mirroring in response to suffering—their brains responded as though they were in pain. Those who had practiced *metta* instead activated brain areas associated with motivation and even reward. Rather than focusing on a victim's pain, they imagined a world in which that suffering diminished.

Some members of the ICN use meditation techniques to turn down their distress. As one nurse describes it, "When you're having an intense conversation with a family, you think about your feet on the ground, and that . . . will stop the cycle of you taking on all their emotions. It separates you a little bit." She supplements those exercises with her own mantra: "This is not my tragedy."

Researchers are now systematically testing the effects of meditation on caring professionals. Across about a dozen studies, physicians who completed meditation-based programs experienced drops in exhaustion and distress. In some cases, they simultaneously reported *greater* empathy in their practice. Eve is connecting these dots. She thinks that by focusing on concern, caregivers can strike a delicate balance between connection and self-protection. At her workshop, residents practice letting go of unhelpful empathy—the kind that might create a heroic need to save patients and lead to shame when they can't. Eve is only now assessing the effects of her program, but other emerging evidence is promising. In one recent study, medical students who practiced a program similar to Eve's reported greater concern for their patients and less depression themselves.

SOME EMPATHY-BUILDING PRACTICES, such as contact with outsiders, are supported by decades of evidence. Work aimed at tuning empathy is much newer. In part, this reflects the nature of the problem. Hatred, callousness, and dehumanization are social emergencies—good people immediately want to combat them. Caring too much isn't as obvious a problem; in some cases, it's revered. Self-sacrificing parents and overworked caregivers can wear burnout like a badge of honor. One social worker told me that she was pulled toward her profession by the "romance of suffering." Combine this with the boot-camp-style toughness of caring work, and you end up with a quiet epidemic.

Caregivers and psychologists are waking up to the fact that they can, and must, wield empathy in more useful ways. We still don't know what will work, when, and for whom. And even the most effective technique might not help caregivers hamstrung by a rushed, impersonal medical system. But this is a burgeoning area of research, with new answers arriving each year. That's good news, given how sorely we need them.

At the deepest level, sustainable, full-contact caring requires professionals to redefine their role. What does it mean to be a healer? To many, it means heroically stepping in to rescue someone from illness, and returning them to safety. At least in the West, medical professionals are our elected champions in this battle to defeat illness and death, a role the ICN staff perform with gusto. Neonatal intensive care is advancing rapidly; survival rates for extremely premature babies have risen steadily over the years. Had she been born in the year I was, Alma might not have survived. "In the hardest times, you can make the biggest difference," Liz tells me. The ICN is one of the saddest places I've ever been, but it's also a miracle factory.

And yet caregivers who think their job is to defy mortality are doomed to fail themselves and their patients. It doesn't have to be this way. "Mortality could and should be part of medicine," Anthony Back tells me. For caregivers, it is an opportunity to help patients not by denying death, but by affirming life. Many of us could take a page from this approach. Empathy sours into guilt and shame when we feel helpless to relieve someone's pain or lessen their struggle. The constant inundation of suffering—on television, online, and in person—might feel like too much to bear, especially when we can't seem to make a difference, but fixing problems is not the only way to demonstrate empathy.

The day after Francisco died, Liz saw grace. She was proud that her team had done everything they could for him and had faced the end with bravery and kindness. "We talk a lot about good deaths and bad deaths here." During college, Melissa volunteered at a hospice center. The work inspired her to go into neonatal intensive care—treating the very young instead of the very old, but still in the space between life and death. "There's so much intimacy and humanity in those times," she tells me. She also describes being with families of dying children as a "privilege," as do at least three ICN nurses.

When doctors, nurses, and social workers manage to be there with families—to listen, explain, and even cry at their side—they give them something irreplaceable. If they can do it in ways that sustain rather than deplete them, they can give that gift to even more people.

CHAPTER 6

Kind Systems

THE PSYCHOLOGIST DAN Batson stressed out seminary students one at a time, and their empathy plummeted. My colleagues and I convinced one person at a time that they could grow their empathy, and in response, they did. Tony McAleer counsels hate group members one at a time, and Raymond Mar studies how fiction helps people understand each other, one reader at a time.

Many of the "nudges" we've encountered build empathy in tightly controlled environments, such as laboratories or counseling sessions. But we don't live in a vacuum; we're part of a larger world, governed by social norms—the beliefs, attitudes, and customs shared by our communities and institutions. Norms affect us in all sorts of ways. People find foods tastier, faces more attractive, and songs catchier when others like them. We litter and vote more often after learning that others have. The extent to which a scandal outrages us, a political candidate invigorates us, or climate change frightens us depends on how people around us feel.

We copy what other people do and think, or at least what we *think* they think. One problem here is that we're often wrong. This is because extreme voices tend to dominate and can be mistaken for majority opinions. Psychologists once interviewed freshmen at Princeton weeks after they got to college, and then again the following spring. They asked two questions: How much do you enjoy binge drinking, and how much does the average Princeton freshman enjoy it? In the fall, students felt lukewarm about it but believed the average freshman was more enthusiastic. This has to be a mistake: Students' own opinions, by definition, make up the *actual* average. But freshmen likely regale their peers with stories about ice luges, not Thursday night study sessions. Loud, unusual opinions crowded out the quieter majority, and students conjured up an imaginary, hard-partying "average" student. By spring, freshmen reported that they enjoyed binge drinking more than they had before. They had invented a norm, then given in to it.

Many of our strongest cultural currents run against empathy. We learn that success requires competition, sometimes even cruelty. As Gordon Gekko put it in *Wall Street,* "Greed clarifies, cuts through, and captures the essence of the evolutionary spirit." This jibes with Darwin's concern about kindness: People who stop to help others won't have the time to innovate, and will inevitably finish last. As we've seen, this is a myth—empathic individuals are *more* likely to succeed in a number of ways. But popular norms have yet to catch up with this insight.

In our polarized era, norms weigh even more heavily against care. As with campus drinking, extreme voices on cable news and social media dominate airspace. They are more partisan than most of us, but they attract so much attention that it's easy to confuse them for majority opinion. Pundits counsel that the other side is an existential threat. Compromising with—or even listening to— outsiders is a form of treason. People conform to this imaginary norm, and it becomes harder to hold on to their own empathy.

A hate group member might be inspired by Tony but soon be surrounded by people who encourage him to rejoin the race war. An ex-prisoner might find hope in a Hemingway story, then go to a job interview where he is reduced to the crime he committed. Prevailing beliefs act on us like gravity; we can escape them momentarily, but more often than not we get pulled back in.

Other norms, though, encourage empathy, and some of these are gaining steam. Many moral revolutions begin when extreme voices demand that we acknowledge one another's experiences. Foot-binding in China and slavery in the United States persisted for centuries, until people came together to abolish them. At the turn of the twenty-first century, few proponents of gay rights dreamed that same-sex marriage would be nationally recognized in the United States. Fifteen years later, it was. And in the fall of 2017, the *New York Times* and the *New Yorker* reported on Harvey Weinstein's history of sexual harassment and assault, spawning new awareness—especially among men—of the fear and pain women endure. Within months, actors had lost contracts, professors were barred from campuses, and Alabama elected its first Democratic senator in twenty-five years, all because people decided they would no longer tolerate abuse.

Conformity gets a bad rap, but in these cases, it led to social change. In laboratory studies, it can spur people to act kindly, for instance, giving to charity or standing up against bigotry. We catch one another's empathy, as well. In a series of experiments run by my own lab, participants read stories about the struggles of homeless individuals. We then showed them how others had responded after reading each story. In fact, these responses were created by us. Half of our participants learned they lived in a caring world, in which their peers empathized a great deal. Half learned that they lived in a callous world, where their peers barely cared at all. Individuals followed suit—reporting greater empathy if their peers had done the same. They then acted upon this feeling: Given the opportunity to donate to a local homeless shelter, people who believed their peers felt great empathy gave more than those who thought their peers were unaffected.

In that study, my lab created empathic and unempathic norms out of whole cloth, and our participants conformed to them. But outside of the lab, we don't need to make anything up. All around us, people act cruelly while others act kindly; people live happily while others are miserable. By focusing on the positive, we can use the force of conformity to pull people toward healthy or kind actions.

The researchers who documented alcohol norms on campus recently did just that. In a newer study, they engaged freshmen in group discussions—revealing that their peers did not *actually* love binge drinking as much as they suspected. Simply pointing out this norm to them reduced students' alcohol use in the year that followed.

The same goes for empathy-positive group attitudes. Civil institutions, HR guidelines, and codes of conduct are all norms in contract form: an agreement to respect one another's experiences and to exclude people who refuse to do so. Empathy is personal, but it's also collective. Organizations that emphasize kindness flourish, even when it comes to the bottom line. In 2012, Google found that its most successful teams were unusually "people oriented": composed of individuals who tuned into one another's feelings and supported each other. The design and consulting firm IDEO encourages employees to set aside time to help colleagues, and considers generosity during hiring and promotion.

Any organization, private or public, large or small, loose or formal, can move in this direction. We are not merely individuals fighting to empathize in a world of cruelty. We are also communities, families, companies, teams, towns, and nations that can build kindness into our culture, turning it into people's first option. We don't just respond to norms; we create them.

THE ONLY GIRL in a family of seven kids, Sue Rahr has never been afraid to tussle. In one of her first patrols with the King County Sheriff's Office near Seattle, she was called to deal with a drunk man harassing customers outside a minimart. "I told him, 'Look, we've got a choice. Either you can go to jail, or you can just go to detox, sleep it off, and everybody's happy.'" She was midsentence when he sucker-punched her. "I immediately reacted not with what I learned at the [police] academy, but what I learned with my brothers. . . . I grabbed him by the hair and pulled him to the ground, and as he's on his way down I kicked him in the nuts."

They wrestled for about a minute before Rahr managed to cuff him, scraping her knuckles on the asphalt in the process. "I never felt the pain because I was so exhilarated by the fight. My supervisor screamed onto the scene, and later said, 'You were standing there with the biggest smile on your face, just dripping blood.'" That night, she told her husband, "Now I understand why boys like to fight; it's fun!"

Rahr also used other skills she'd picked up around her brothers in her policing. "Growing up with them, I had to learn to maneuver, outsmart, and influence." In King County, new cops spend their first three months with a field-training officer (FTO) who ensures they can fulfill their duties, including combat. Sue was the only officer who made it through that time without ever needing to use force (this was before her minimart encounter). Her FTO left that part of her assessment blank.

Rahr now heads police training for the entire state of Washington. Over a half dozen years, she's built a new system with new expectations for cops, hoping to bring empathy back to a profession that some feel has lost its way.

Modern policing is a surprisingly young line of work. Two centuries ago, patchwork forces settled disputes and punished crimes, even in London. By the 1820s, it became apparent that London required more organized law enforcement, but many of its citizens disagreed. They envisioned military forces stomping through their streets, ready to strip away their liberties. The task of easing their fears fell to Sir Robert Peel, Britain's home secretary. Peel was a brilliant scholar and a canny politician—he'd later serve two terms as Britain's prime minister. He realized that a police force could succeed only if it had citizens' cooperation and inspired their trust.

In 1829, Peel introduced the Metropolitan Police Act, calling for a force of several hundred constables (they were, and still are, called "bobbies" in his honor). Bobbies lived restricted lives. They worked seven days a week, couldn't vote, and had to ask approval to marry or even have a meal with citizens. They were required to wear their uniform—navy-blue tails and top hats—even when off duty, to assure citizens they were not being spied on. Along with the act, Peel laid out a vision of policing that today reads like an idealistic fantasy. "The power of the police," he wrote, "is dependent on public approval of their existence, actions, and behaviour." He demanded that officers "use physical force only when the exercise of persuasion, advice, and warning is found to be insufficient." Most famously, he wrote, "The police are the public and the public are the police."

As policing made its way across the Atlantic, Peel's ideas came with it. American police officers typically lived in the communities they patrolled. They arrested thieves, but also operated soup kitchens and helped immigrants find work. They were rewarded based not on the number of arrests they made, but on their ability to ensure an orderly beat. In the twentieth century, cops became less neighborly and more professional, but they still upheld cooperation as a core part of their job. Officers around the country pledged their commitment to "community policing": a vague but warm notion that found them playing pickup basketball games and attending bake sales.

In recent decades, these ideals have eroded, in part in response to escalating violence. As the drug trade grew, criminals amassed weapons and were in some cases better armed than the police. In 1965, a routine traffic stop in Los Angeles exploded into a six-day riot; thirty-four people died and more than a thousand were injured. The following year, Charles Whitman lugged eight guns and seven hundred rounds of ammunition to the top of a University of Texas clock tower. Over the next hour and a half, he shot forty-four people, killing thirteen.

To many, it felt like America's streets were becoming war zones.

By the 1970s, about two officers were gunned down in the line of duty each week. At the same time, the police created their own battalions: special weapons and tactics (SWAT) units. SWAT teams were intended only for extreme situations, such as armed bank robberies, but their use soared: In 1980, SWAT teams were deployed about three thousand times; by 1995 this number had risen to thirty thousand, though the crime rate remained steady. In 1996, President Clinton signed into law that year's National Defense Authorization Act. It included the 1033 program, under which police departments could request surplus equipment from the Department of Defense. By 2014, over $4 billion worth of hardware had streamed through the program. American policing entered the age of armored vehicles, matte black body armor, and assault rifles.

Along with military equipment, a new philosophy took root among American cops. The "warrior mentality" encouraged police officers to view themselves as combatants embedded in dangerous communities. This ideology spread quickly and appealed to many. It valorized cops' bravery and honored the risks they take. It allowed them to band together like soldiers against a common enemy. But it also turned every non-officer into a threat. Cops across the country were conditioned to expect danger around every corner. In 2014, course materials from a New Mexico police training facility were obtained by the press. They instructed cadets that during every routine traffic stop they should "always assume that the violator and all the occupants of the vehicle are armed."

No one encapsulates warrior philosophy better than Dave Grossman, the nation's most prolific police trainer. Grossman delivers "The Bulletproof Warrior"—a frenetic, six-hour-long seminar—upward of two hundred times a year to new police officers, veteran cops, and groups of "armed citizens." He paces across stages around the country, describing America as a violent fever dream. "The number of dead cops has exploded like nothing we've ever seen," he pronounces, as though every audience member has a target on their back. In fact, the opposite is true: a police officer serving in the 1970s would be more than twice as likely to be killed in the line of duty than one serving in the past decade.

According to Grossman, the only way for cops to survive in this frightening world is to become frightening themselves, by always being prepared to use deadly force. He calms audiences' apprehension about legal blowback. ("Don't be afraid of being sued," Grossman tells them. "Everybody gets sued. It's just a chance for overtime.") He tells them that the night after their first kill, they'll have the best sex of their lives. And he warns them that if they are *not* ready to kill a murderer, his victims' blood will be on their hands. Grossman wants police officers to turn killing into a reflex. Much of their training is already geared to doing just that: Three-quarters of police officers never fire their weapon in an entire career, but they nonetheless spend hundreds of hours firing at paper targets in the academy.

The warrior mentality places cops in a psychological powder keg. They come to believe that their only option is to dominate citizens rather than listen to them. Seth Stoughton, a professor of law and former police officer, describes this bind: "If I'm worried about never making it home again, I don't really give a damn if I offend someone. Whatever emotional toll my actions take on them, it will feel less important than my survival." Fear and anxiety also make violence more likely. Psychologists have demonstrated this through an unfun video game called the "weapons identification task." Players watch as scenes flash across a screen: a schoolyard, a street corner, a park. In the middle of each, a man—either black or white—holds either a phone or a gun. If the man is armed, players press a key to "shoot" him; if he is unarmed, they press another to not shoot. Players are quicker to shoot black targets holding a gun and more likely to mistakenly shoot unarmed black targets. Under stress, they grow even more trigger-happy and more racially biased.

Warrior policing shreds Peel's principles. It also makes it harder for individual cops to go against the grain. A young, idealistic officer might want to help citizens, but in a warrior culture, that attitude would be mocked as dangerously naïve. Trainers will warn their new colleague that he's surrounded by criminals, whether he wants to believe it or not. Odds are that eventually he'll mold himself to fit the culture around him.

Police officers are safer than they have been in decades, but coming into contact with them is more dangerous. In the United States in 2017, almost five civilians were killed by police officers per day, more than twice as many as in the year 2000. Thanks to ubiquitous recording, this violence is more visible than ever. Over and over again, the nation sees another black or brown face for the first and last time as an unarmed citizen dies at police hands. This has led to a two-decade low point for public confidence in law enforcement, and for race relations more broadly.

Most cops just want to do their jobs and return home to their families; so do most of the citizens they pull over. But the distance between officers and the communities they've sworn to protect has never seemed greater.

RAHR WORKED IN the King County Sheriff's Office for thirty-three years. She served in every unit, from sex crimes to gang violence, but spent her formative time in internal investigations. She encountered dozens of police misconduct cases, and after a while she found it hard to believe that every culprit was a rotten human being. Many of them had inherited their instincts from a broken culture. "I thought, rather than focus on bad apples, let's think about the barrel." In 2012, she took over as executive director of the Washington State Criminal Justice Training Commission (CJTC). Every law enforcement officer in the state comes through CJTC; by now, more than three thousand have trained under Rahr.

CJTC's woodsy headquarters in Burien, Washington, recalls a college campus, at least if you can ignore the students marching in formation. The walls are covered with pictures of every CJTC recruit class. Members of class 1a, from 1938, look like Humphrey

Bogart understudies from *Casablanca*. Rahr looks a little uncomfortable in her class 114 picture, from 1979. The week after I visit, CJTC will graduate class 735. Police officers spend nineteen weeks at CJTC; corrections officers, four. The training is relentless. As we stroll the grounds, a supervisor tells me about a recruit whose wife is being induced on Sunday. "I assume he'll take Monday off."

Much of CJTC's curriculum is standard issue. Recruits spend 120 hours on defensive tactics, practicing baton techniques on muscular mannequins and sparring partners. In the shooting range, officers slowly pace sideways while firing at posters of stereotypical-looking criminals. Altogether, CJTC recruits fire about a million rounds a year. After each drill a white-mustached training officer retrieves spent shells using a specialized, caged cart like you might see scooping up golf balls at a driving range.

That's where the similarities between CJTC and typical police training end. Above the academy's entrance, a sign reads IN THESE HALLS, TRAINING THE GUARDIANS OF DEMOCRACY. This is meant to remind recruits of Rahr's most important mandate: that they reject the warrior mentality and instead see themselves as caretakers of their community, working with citizens to keep everyone safe.

On every desk in every CJTC classroom, another motto is written on a folded, laminated card: LEED: LISTEN AND EXPLAIN WITH EQUITY AND DIGNITY. For decades, the psychologist Tom Tyler has demonstrated that powerful people—doctors with patients, police encountering citizens—garner respect when they are transparent, impartial, and attentive, even while delivering punishment. "I've had lots of people thank me for arresting them," Rahr says, "or at least for being decent with them while I did it." LEED is her encapsulation of Tyler's ideas; she calls it "a Happy Meal version of a research smorgasbord."

Guardianship is a poetic but fuzzy idea. On the ground in Burien, Rahr and her staff make it concrete, in three ways. The first is by example. Prior to Rahr's arrival, CJTC operated like a boot camp. Drill sergeants broke down recruits and built them back up.

The first time Rahr walked the halls, recruits snapped to attention as she passed. She found it startling, and useless. "We don't need cops to salute," she thought. "We need cops to talk." Rahr stripped away the military style in favor of a more open atmosphere. "If the organization itself, as a culture, isn't procedurally just to [recruits], then they're more likely to go out into the field frustrated, thinking, 'Well, this is all bullshit.'"

The second is classroom instruction, or as teacher Joe Winters affectionately calls it, "death by PowerPoint." Recruits take classes on emotional intelligence, "heart math," racial bias, and mental illness. They discuss how to tell whether someone who's publicly naked is suffering from a manic episode or on a methamphetamine binge, and they practice talking people down from suicide and delusions. Lecturers remind recruits that when people commit crimes, they're often in deep distress. "You're seeing them in many cases on the worst day of their life," Rahr explains. "They're going to be acting like jerks, but that's because of their situation."

The third arm of guardian training takes place in "Mock City," a gymnasium converted into fake stores and apartments, with most furniture replaced by foam boxes. It feels like a low-rent movie set, complete with actors who play criminals and victims. Here, recruits strap on wooden guns and practice managing volatile situations until they get it right.

The day I visit, Mock City is full of recruits who had failed their last session. If they fail again, they won't graduate with the rest of their class. We head to a simulation in which a father (Joe Winters plays the part) is standing near a building. Two recruits arrive, and he tells them about his son, who is suffering a psychotic episode inside. "I'm afraid he'll hurt himself, or me," Winters says. The recruits burst in and find a young man sitting atop a bed in a room full of toppled foam boxes, holding a baseball bat and talking to voices in his head. The acting is not top-notch. "Stop talking to me!" he yells between bouts of mumbling and rocking, trying valiantly to capture the made-for-TV-movie version of schizophrenia.

The recruits interview him, convince him to relinquish the bat, and apologetically cuff him.

Outside, Winters—no longer pretending to be the young actor's father—asks what the plan is. "Take him to the hospital," one recruit responds.

"On what grounds?" Winters counters. The recruit mentions the bat and broken property, but Winters isn't buying it. "It's not illegal to have a bat, or to break your own stuff. What did I *say* to you before you went in?"

The recruit freezes. He's terrified, because he can't remember the fear Winters described when playing the boy's father. Without that, he has no legal cause for apprehending the son. As one trainer puts it, "Crimes need a victim, and lots of times victimhood depends on a person *feeling* threatened."

It's startling to see a future police officer this upset about failing to pick up on someone else's feelings. But at CJTC, tuning into emotions is a core part of police training, and not only for the purpose of establishing grounds for arrest. This reflects Rahr's viewpoint: "In law enforcement," she explains, "empathy is still viewed as a weakness, or catering to political correctness, but really it's critical to officer safety. Police officers deal with people in crisis, and having your trauma acknowledged lowers the tension. Listening is a de-escalation strategy."

CJTC'S TRAINING HAS some important flaws. Trainers seldom consult with psychologists, and sometimes it shows. Recruits learn about their Myers-Briggs personality types, though that test is scarcely backed by research evidence. During a mental illness workshop, a trainer spent much of his time describing "excited delirium." This drug-induced state renders people aggressive, insensitive to pain, and unnaturally strong—in other words, Hulk-like. I had never heard of this condition, and later learned that it is not recognized by the American Psychiatric Association, but is often—and controversially—cited by cops to justify the use of force.

Nor does CJTC do much to educate recruits about race. They discuss bias in the classroom, but even Rahr describes that session as "a little antiseptic." Race-based drills at Mock City have proven logistically difficult, she says, because of the lack of diversity in their pool of actors, but that is a thin rationale. Rahr once organized a discussion between recruits and a black community activist, which quickly devolved into a shouting match. Perhaps that means this is exactly the conversation more recruits need to be having.

Nonetheless, CJTC bakes empathy into police culture. In its ecosystem, professional success is tied to cooperation with citizens—a photonegative of Dave Grossman's approach and a return to Peel's principles. A warrior cop can't do her job if she's not ready to gun citizens down. A CJTC recruit can't become a cop until she's ready to hear them out.

Washington State is home to more than three hundred sheriff's offices and police departments. Each has its own culture, and Rahr acknowledges that many are the opposite of hers. She imagines many of her recruits move to their new departments and meet field training officers who tell them to "get past that touchy-feely bullshit and do real policing." But early research suggests that CJTC's approach makes a difference even after cadets leave Burien. Its graduates report greater empathy than other cops, but they also show more care in their policing.

In a recent study, psychologists selected three hundred Seattle officers who worked in high-risk areas and ran them through LEED training. In the subsequent months, these officers used force 30 percent less often than their peers. CJTC's approach to dealing with mental illness has also gained traction. In the past three years, Washington cops have shifted their tactics around mentally ill individuals: arresting fewer and hospitalizing more.

In the wake of Michael Brown's 2014 death at the hands of

police in Ferguson, Missouri, President Obama convened the Task Force on 21st Century Policing. Rahr was part of the task force, and her philosophy was adopted in its final report: "Law enforcement culture should embrace a guardian—rather than a warrior—mindset," it reads.

Guardianship values have slowly spread across the country. In 2017, LAPD chief Charlie Beck released a new policy demanding that his officers should "always be guided by compassion and empathy in all of their interactions" with homeless people. CJTC-style training has caught hold in Las Vegas; Stockton, California; and Cincinnati. Decatur, Georgia's police recruitment video opens with its chief declaring, "We're a very empathic department. We try our very hardest to put ourselves in other people's shoes."

Norms are changing, but it's impossible to tell how much of a dent this will make in American policing writ large. For each volley of praise Rahr and CJTC receive, they lose support from skeptical peers. Twenty percent of CJTC's senior staff turned over after Rahr introduced the guardian philosophy. In 2016, she spoke at the FBI National Academy's annual conference; many regular attendees boycotted. Ozzie Knezovich, Spokane County's sheriff, has been troubled by the officers he sees coming out of Rahr's CJTC. "My field training officers tell me, 'It's scary, Sheriff. [CJTC recruits] will not engage, and it's going to get one of them killed.'"

Knezovich was trained in Wyoming, at a college-style academy without military airs. "I was never taught to be a warrior," he tells me. Knezovich thinks Rahr merely rebranded standard police ideals in prettier language and then took credit for them. Worse, he thinks that by raising the specter of warrior policing she has deepened the divide between police and communities. "The biggest challenge a chief has today is to convince law enforcement officers that the community cares about them, and then to convince the community that law enforcement cares about them." When Rahr paints most American policing as a street war, "that decimates the public trust."

Knezovich is passionate, but it is odd to lay the fear of cops at Rahr's feet. Communities of color, especially, need no one's help to see the police as a militaristic, volatile force. That needs to change one interaction at a time, but it requires a change in the culture. Consider Antonio Zambrano-Montes, a thirty-five-year-old father of two who is battling mental illness. In early 2015, Zambrano-Montes started throwing rocks at passing cars in Pasco, a small city in southeast Washington. When three police officers arrived, he threw rocks at them as well but was otherwise unarmed. In a video of the event, he runs across the street, fleeing from the officers. He is still running when they fire, killing him.

The video, like so many, seems like all the evidence one would need to prosecute the officers. Unlike most states, however, Washington required that an indictment prove an officer not only behaved violently but did so with "malice." Barring telepathy, this rule makes it nearly impossible to convict police officers for excessive use of force. During my visit to CJTC, the officers in Zambrano-Montes's shooting were cleared of all charges. Rahr supported the decision, adding that "the officers involved, as individuals, are really good guys." She reminded me that most officers who kill do so inadvertently. "We generally don't prosecute people for making an honest mistake. . . . Prosecution is such an extreme response to bad judgment."

Here Rahr is uncharacteristically disconnected from the public's perspective. As a society, we *do* prosecute people for involuntary manslaughter and criminal neglect. And the term "honest mistake" is better suited to a mixed-up order at a restaurant than the killing of a civilian. But Rahr's views on this are in line with those of almost every officer I meet at CJTC. They want me to know that most cops are nothing like the violent racists you see on YouTube or Periscope. They want me to know about the risks they take to protect even people who would like to see them dead. They have a point. Police and citizens come into contact hundreds of thousands of times a year, and the vast majority of these encounters end peacefully. Videos of police violence overshadow the goodwill of more than a million public servants.

But in another sense, it doesn't matter what percentage of encounters go right, when the ones that go wrong look like executions. Rahr concedes that "the optics are terrible." Other officers also mention optics but quickly add that those perceptions are wrong: based on biased reporting that throws cops under the bus.

This highlights a major tension at the heart of CJTC's mission. Rahr and her team encourage kindness among cadets who, as officers, will face few consequences if they act cruelly. Peel's principles hold that the police and citizens should join together in a single community. Training officers to befriend, listen, and operate fairly is an obvious step in that direction. But what does empathy mean without accountability? When an officer can kill you with legal impunity, having them ask how you feel is cold comfort.

Part of the problem is that even police officers who empathize with citizens often empathize more with one another. In a 2017 poll, 60 percent of the public thought that officer-involved shootings represented a broad problem with police culture. More than two-thirds of officers disagreed, calling them isolated incidents. When things go wrong, cops often circle the wagons.

Emile Bruneau, who works to build empathy in the face of conflict, has examined this sort of "empathy bias." He recently asked Americans, Hungarians, and Greeks how they felt about their own group compared to outsiders they've historically disliked (Arabs, Muslim refugees, and Germans, respectively). He also asked about their willingness to cooperate across group lines. Highly empathic individuals didn't necessarily support peaceful policies, especially if they cared more for their own group than outsiders.

This work has a surprising implication: Sometimes compromise is best served not by building empathy for outsiders, but by *reducing* empathy for insiders. This is a tall order for any group, but especially for cops. Doubting one's friends is painful, and it is dangerous when you depend on them in high-risk situations. But in order for police officers to repair their relationships with wary communities, they may need to treat their colleagues with more skepticism, acknowledging wrongdoing even when it involves people they admire. Enforcing that norm could move us toward what Rahr describes as her ideal world, one in which "everyone who sees a police officer would have the initial reaction: 'I'm safer.'"

A FEW MINUTES after he arrived in jail, Jason Okonofua was told to strip naked. When he got down to his underwear, he hesitated. He was sixteen years old. An hour before, he had been in his tenth-grade AP calculus class. This didn't seem right. "I said take off your clothes!" the guard shouted.

Jason was in juvenile detention for only a few hours, but fifteen years later he remembers it vividly. "It implanted something inside me. . . . 'You're nothing, you have no rights whatsoever.'" For years he had done everything right: Jason was a straight-A student and a middle linebacker on the football team. He was poor and black, but he was working to make his own luck. Jail dissolved his confidence in minutes. "It was a feeling like no, none of that even matters; you still don't have control over your life."

Jason's father and mother were born in Nigeria and Knoxville, respectively. "How they ended up together," he tells me, "I have no idea." They divorced when he was five, and the split hit his two older brothers hard. The oldest was nine at the time of the divorce and had been fast-tracked to a gifted and talented program. A decade later he graduated high school with a GPA of 0.57.

Jason did his best to avoid trouble, but he admired his brothers and wanted to be loyal to his family. Sometimes that meant backing them up in brawls or hanging out in the wrong circles. "Those few times could have changed my entire life if the police happened to show up at a certain time, or if I didn't run when the police *did* show up." One of them joined a gang, and both older brothers were expelled from schools all over Memphis. Jason was uprooted alongside them. Each school they landed in was poorer, blacker, and more prison-like than the last. By tenth grade Jason had attended seven, and was used to walking through metal detectors to get to class.

He had watched his brothers follow a clear path. "The first time they got in trouble fed into the next time," Jason remembers, "and slowly but surely they stopped caring about getting in trouble." School staff had often heard about the Okonofuas by the time they arrived, and their expectations laid track for the brothers to fail. "Teachers . . . would think 'Oh, you're a bad kid; as soon as you do something wrong we're kicking you out. We don't want you here.'" Jason aspired to do better, but in many teachers' eyes, belonging to his family limited who he was allowed to be.

In the fall of his sophomore year, Jason was sitting at lunch when a senior handed him a flyer for a party. It read, "BYOBB: Bring Your Own Beer and Bitches." Before Jason even knew what it was, a vice principal rounded up anyone holding a flyer and brought them to her office, where she handed out one-day suspensions. Jason looked around and thought, "I'm not one of these kids." When his turn came he refused the suspension, adding that he needed to get back to studying. "Oh, okay, you're going to be insubordinate?" the vice principal replied. "That will be a three-day suspension then." Jason refused again. She called the school police officer and had him arrested for disturbing the peace.

The officer took Jason to a back room in the principal's office, cuffed him, and began his report. They were alone. Like most people at the school, he knew Jason's brothers. He looked at Jason and said, "I thought you were the good one."

IN 2011, ABOUT three and a half million children were suspended from school, at more than double the rate recorded in 1975. "Exclusionary discipline," which includes suspension and expulsion, skyrocketed in 1994 with the Gun-Free Schools Act, which mandated that any student caught with a firearm at school be suspended for one year. Over the years, "zero tolerance" policies spread beyond guns to knives, drugs, and threatening behaviors. Then the definition of "threatening" got fuzzy. Kids were suspended or expelled for chewing Pop-Tarts into the shape of guns, or taking "drugs" such as birth control.

Zero tolerance policies are education's version of warrior policing: norms that are meant to promote order but create animus instead. In schools, these policies are meant to deter students from dangerous and illegal behavior, but there's little evidence they do; in fact, after being suspended, students are more likely to act out, drop out, and be arrested. Even students who are not suspended suffer when their peers are—their standardized test scores drop, they trust teachers and principals less, and they grow apathetic and anxious. By excluding students, schools create more of the chaos they are trying to prevent. Policy makers rolled back zero tolerance policies in 2014, but in response to an epidemic of school shootings, the Trump administration has recommended bringing them back in full force, though there's no evidence that suspensions curb violence.

Black and brown children are three times more likely to be suspended than their white peers. Some cases reflect students' real misbehavior, and some reflect teachers' racism. In other cases, the problem lies with a culture that sets traps for students and teachers alike.

New teachers often describe themselves as "idealists"; within this group, over 40 percent specifically hope to create opportunities for underprivileged and minority students. Many of them become disheartened by the difficulty inherent in controlling unruly students; they are forced to become disciplinarians, and this is a complex role. Troublesome students are often troubled—by stress at home, bullying, or low self-esteem. Teachers with the time, interest, and energy can use disciplinary action as an opportunity to check in with students about what went wrong, listen to their perspective, and offer support. Assuming a given student will be there all year, helping them is probably the best way to help the rest of the class as well.

Zero tolerance culture reverses this norm. A teacher's job no longer rests on understanding difficult students. Instead, they must protect classrooms from dangerous elements. This encourages them to identify "bad kids" early and react to them forcefully, expunging threats to the system like an antibody hunting down a virus. Like warrior policing, zero tolerance turns people who could have worked together into adversaries.

These pressures mix with common racial stereotypes to form a toxic cocktail. White students are suspended more often for concrete infractions such as possession of cigarettes, but black and brown students are suspended more often for vague transgressions—such as "disrespect"—which depend on teachers' judgment. When Jason refused his suspension, his vice principal did not see a reasonable challenge. She saw a disturbance of the peace.

Kids learn from these norms, and become who we expect them to be. Minority students often suspect that teachers are prejudiced against them. Unfair discipline confirms this and makes school feel like a courtroom. When students feel disrespected, they misbehave more; teachers in turn feel more threatened and escalate their discipline. Each learns to fear and provoke the other in a cycle that spins like water around a drain and eventually flushes thousands of students out of school altogether, often into the arms of law enforcement. Jason fought against this throughout his childhood, but he was swimming upstream.

The day after his visit to juvenile hall, Jason and his mother went to court, where his fate was put in the hands of a judge. The school had pressed charges, which would have marked the beginning of Jason's criminal record. But as the judge leafed through Jason's file, he paused. "Are these *all* honors classes?" he asked. They were. "And you're getting As in *all* of them?" Jason was. The judge paused again, and then said, "Get out of here and tell your school to never send you back." Jason's record was expunged; legally speaking, his arrest never happened.

That summer, the Memphis school system arranged for a group of high-achieving students of color to spend their summer at elite prep schools, and Jason was sent to St. George's in Rhode Island. He so excelled there that by the time he returned to Tennessee, the prep school had offered him the chance to come back for the academic year on a full scholarship. He spent his last two years of high school at St. George's. It was a different world. He was used to schools with security cameras in the hallways, but this one had a private beach, in-house college advisers, and a teacher for every eight students. His roommate's mother—the doyenne of an aristocratic Providence family—brought cookies every week.

What most startled Jason was not how different his circumstances were, but how differently he was viewed at St. George's. "In the Memphis school . . . standing up for myself led to me getting arrested," he tells me, while "those same things at the prep school were encouraged." When Jason challenged teachers, they suggested that he join the mock trial and the debate team. Jason flourished at St. George's, and then as an undergraduate at Northwestern and a PhD student in Stanford's psychology program. He's now a professor at UC Berkeley.

His research examines the type of unfair discipline he once encountered. In one study, Jason asked teachers to read about an imaginary student who misbehaves in class. When teachers believed the child was white, they were more willing to strategize about ways to help him. When they believed he was black, they were more likely to say they'd suspend him.

Jason didn't just want to study racial disparities in education; he wanted to combat them. As a graduate student, he joined a core of researchers at Stanford who were working on this problem. They had found that minority students tend to feel unwelcome at school, especially in predominantly affluent, white settings like Stanford. They pioneered an intervention in which new freshmen spent time reflecting on why they *do* belong at Stanford. Remarkably, this simple exercise cut the racial gap in students' GPA in half over the subsequent year.

Jason wanted to use this work to improve the experiences of minority students in high schools. In doing so, he joined a massive educational trend: focusing on students' feelings. Social and Emotional Learning (SEL) programs teach children how to regulate themselves and care for one another. SEL takes many forms: Students might practice mindfulness, start each day by naming their emotions, or talk about the way their actions affect other kids. Dozens of schools have adopted a full "kindness curriculum" developed at the University of Wisconsin's Center for Healthy Minds.

A recent review of more than two hundred studies—including about a quarter of a million students—demonstrates that SEL programs make a difference. After participating in them, students better understand one another's feelings and control their own moods. Other benefits are more tangible. SEL programs decrease bullying, depression, and disciplinary troubles, and raise GPAs.

Their biggest weakness: SEL programs work less well in older kids. By the time they're teenagers, students are almost entirely immune to them. This could happen for any number of reasons: Teens face the chaos of puberty and increasing academic pressures as they prepare for college. They also become more attuned to their social world. Adolescents conform to each other more than any other age group, and if other students don't care—or, worse, think kindness is for dorks—working on it becomes social suicide.

Consider Drug Abuse Resistance Education (DARE). In DARE sessions, cops come to classrooms and show students pictures or even samples of drugs. The officer warns them that their peers will think drug use is cool and pressure them to join in. The punch line: Doing the right thing means *not* joining the in crowd. This is a fine message on its face, but it often backfires. It highlights dangerous norms and asks students to fight against them—but as we've seen, norms tend to win. Despite costing tens of millions of dollars,

DARE does not appear to have reduced drug abuse among children, and some evidence suggests it's made things worse.

A smarter strategy would be to work *with* norms, not against them. Betsy Levy Paluck recently used this approach in fifty-six New Jersey middle schools. Paluck deputized groups of kids to identify their campuses' worst social problems, such as bullying and rumor spreading. They then created campaigns, slogans, and posters encouraging kindness, and papered their schools with them. Where other anti-bullying programs have failed, this approach took off. Disciplinary problems plunged, and students reported that their peers cared more about one another. Instead of fighting against conformity, Paluck used it to build healthier environments.

My graduate student Erika Weisz and I are taking a similar approach to building empathy in teenagers. Erika and her team have worked with about a thousand seventh graders around the Bay Area. They first ask students to write about why they think empathy is important and useful. Next, students read one another's messages, learning that their peers value caring as much as they do. They also read empathy-positive messages written by Stanford students—a group that kids in the Bay Area tend to admire—as part of Erika's earlier study of empathic mindsets. Finally, we asked them to imagine talking to a student from another school and bragging about how empathic their class was.

Among angsty adolescents, it's easy for bullies, social climbers, and mean girls to dominate conversations. Like hard-drinking college freshmen or cable news anchors, these extreme voices can crowd out the majority. Erika's approach is to help students notice that the majority of their peers *do* care, giving them the opportunity to conform to an empathic norm. Though our data are still preliminary, her efforts seem to be working: after learning about their peers' empathy, students told us they were more motivated to empathize as well. That motivation, in turn, predicted how kindly they acted toward their classmates.

In building his intervention, Jason also focused on classroom norms. But instead of putting the onus on students, he decided to focus on teachers. From his days in Memphis, he knew how much teachers matter, especially when students struggle. If he could encourage teachers to empathize with students during tough moments, he thought, perhaps their relationships would go down a better path.

He set up shop at five Bay Area middle schools, training math teachers at each one to deliver "empathic discipline." Teachers first read about the reasons that even good kids act out—such as insecurity and puberty. They then reflected on discipline as an opportunity not just to punish students, but to help them grow. Jason provided stories from students who described the difference receiving empathic discipline had made for them. One note read:

> One day I got detention, and instead of just sitting there, my teacher talked with me about what happened. He really listened to me. . . . It felt good to know I had someone I could trust in school.

Jason then asked teachers to write about their strategies for disciplining kids. Teachers responded by extolling the virtues of kindness. One wrote, "[I] greet every student at the door with a smile every day no matter what has occurred the day before." Another, "I NEVER hold grudges. I try to remember that [students] are all the son or daughter of someone who loves them. . . . They are the light of someone's life!"

These attitudes were reflected in the classroom. After teachers learned about empathic discipline, their students reported feeling more respected. This was especially true of students who had previously been suspended. Rather than making them feel like outcasts, empathic teachers created an environment in which they could thrive. And thrive they did. Students whose math teachers received Jason's training were suspended about half as often as those in other classes. This difference was again strongest for adolescents most likely to struggle: boys, African Americans and Latinos, and kids who had been suspended in the past.

Jason's work is preliminary but powerful—in part because its effects can't be boiled down to teachers' empathy alone. "[The intervention] was only with math teachers," Jason explains, "but students were less likely to get suspended by *any* teacher, in the hallway, on the playground, on the bus ride home." Other teachers didn't know which students had received empathic discipline and which ones hadn't. Instead, it seemed that students whose teachers empathized with them behaved better not only in that class but elsewhere as well.

Schools around the world encourage students to adopt a growth mindset—believing in their capacity to change for the better. But Jason's work shows us that sometimes—especially for struggling students—self-perception is not enough. "Mindsets are in a person, and that's important," he reflects, "but we should also think about where those mindsets come from." People choose how to interpret their environment, but we also create environments together. These circumstances shape what we expect of one another and of ourselves.

Those of us in power have a responsibility not only to be kind but also to create ecosystems in which kindness is expected and rewarded. Schools, police departments, families, companies, and even governments that take this approach make empathy easier for the people within them.

Jason is doing his part, and more. His teacher training has exploded; it is now being used in school districts in Florida, Pennsylvania, Virginia, and Georgia. Someday soon it may reach Tennessee and the vice principal who once sent Jason to jail. Perhaps it can help her think differently about the next child who crosses her path.

CHAPTER 7

The Digital Double Edge

IN 2007, THE Iraqi artist Wafaa Bilal invited anyone in the world with an Internet connection to shoot him. Three years earlier his brother Haji had been killed by a drone-directed missile strike in their hometown of Najaf. Grieving, Bilal saw an interview with a drone "pilot" who dropped missiles on Iraq from a base in rural Colorado. She cheerily described her job, making killing sound less emotionally troubling than most video games. "It struck me," Bilal recalls, "that Haji's death had been orchestrated by someone just like this young woman, pressing buttons from thousands of miles away . . . oblivious to the terror and destruction they were causing to a family."

He wanted to hate this pilot but also knew she wasn't cruel in any special way. Violence mediated by technology doesn't feel violent, because people mediated by technology don't feel like people. Bilal decided to open an exhibition titled *Domestic Tension* (he preferred the title "Shoot an Iraqi," but his gallery hosts wouldn't allow it). He set up a spartan, all-white living space in Chicago's Flatfile Gallery with nothing but a bed, a small computer desk, and a paintball gun. It was mounted on a small turret that could be controlled by remote users, with a webcam affixed to its barrel. (Bilal got the idea from a website that allowed people to hunt penned-in animals via remote-controlled firearms.) Users could log on, aim, and fire as many times as they liked. Bilal switched it on and stayed in the room for the following thirty days.

During that time, the gun went off about sixty thousand times, or about once every forty-five seconds, twenty-four hours a day. Bilal took shelter behind a Plexiglas barrier while he slept, but sleeping was nearly impossible until he grew accustomed to the noise, or at least exhausted enough to nap through it. A few days into his experiment, Bilal's room looked like a canary-colored Jackson Pollock painting. Dozens of shooters vied for control of the gun at the same time. Online traffic crashed its servers. People shot at Bilal from 138 countries, often while mocking him through chat messages. "Die, terrorist," read one.

Some people would have shot (and yelled) at Bilal while standing in front of him. But he contends that many of his attackers fired only from the safety—and emotional distance—of the Internet. By now, this idea is so obvious as to be quaint. Technology is widely viewed as our era's biggest threat to empathy.

IN THE LEAD-UP to the 1964 World's Fair, Isaac Asimov was asked by the *New York Times* to imagine what that same event might be like in fifty years. His first guess was that the use of "electroluminescent panels" would become ubiquitous. Through them, he wrote, "men will continue to withdraw from nature in order to create an environment that will suit them better." Since *Domestic Tension* closed, the world has been introduced to Instagram, Uber, Airbnb, Kickstarter, Bitcoin, Tinder, and the iPhone. Finance, friendship, romance, and identity are all brokered online, and the online world is with us constantly. In 2007, the average American adult spent eighteen minutes a day on their phone. By 2017, that had ballooned to four hours a day.

Asimov's future is our present. This worries many people, who warn that technology will leave us dumber, sadder, and meaner than we were before. And going digital *does* have psychological costs. In one experiment, my colleagues and I asked people to take a tour of Stanford's Memorial Church. We temporarily confiscated some of their phones, and required others to take pictures of the tour and post them to Facebook. Those who shared their experiences online ironically remembered them less. People's attention also shrinks amid the bottomless scrolling feeds of social media platforms.

Robert Vischer coined the term *Einfühlung,* or "feeling into"—empathy's linguistic predecessor. He was not a psychologist but an art theorist, and he thought of *Einfühlung* as a state of close attention, which allows viewers to truly "see" the emotional meaning behind sculptures and paintings. If human connection, likewise, is a process of truly seeing someone and being seen in return, we should celebrate the Internet. Through it, we can access the lives of millions of people, in every country, on their own terms, and broadcast our lives back to them. News outlets, publishers, and other traditional gatekeepers no longer dictate who or what we care about. Technology takes untethering and explodes it, all over the world, all the time.

The philosopher Peter Singer believes our "circle of care" has expanded over time, slowly overrunning the boundaries of family, village, and country. The Internet should finally widen this circle to embrace all of humanity. In a cheery 2014 op-ed, Mark Zuckerberg said as much. "Perhaps the most important change" social media can create, he wrote, "might be a new global sense of community. Today we can only hear the voices and witness the imaginations of one-third of the world's people. . . . Tomorrow, if we succeed, the Internet will truly represent everyone."

A few years later, Zuckerberg's predictions seem almost comically off base. What went wrong?

Technology allows us to "see" an unprecedented number of people, but what we get back is thin gruel compared to old-fashioned social contact. Real-world conversation is rich and multifaceted; we catch a glint in a friend's eye when he talks about his last date, a hesitation in his voice when he says work is great. We see excitement and hear doubt. Emotions are palpable and easy to share. The more time we spend with people, the better we become at reading them, and the more we care about what they feel. Online, social life is reduced to strings of text and images. And especially for younger generations, those interactions are increasingly replacing analog hangouts.

By retreating from face-to-face encounters, we have neglected the world's best naturally occurring empathy training. Has this reduced our ability to connect? It's hard to tell. Empathy has declined over the past thirty years while technology use has increased, but the fact that two things co-occur does not mean one produced the other. But there are other troubling correlations: Countries with greater Internet usage also have lower average levels of empathy, and individuals who spend relatively more time on the Internet, social media, or gaming platforms report greater trouble understanding others. And when people read each other's words, as opposed to hearing their voices, they're more likely to dehumanize them—especially if they disagree with what they say.

Even though we *can* see anyone online, we often use that power to narrow, not broaden, our perspective. Inundated with more stories and statistics than one could ever hope to process, we must choose where to point our attention. That opportunity feeds our laziest psychological impulses. We seek out facts that match what we already believe, and we encase ourselves in echo chambers of the like-minded. We also gravitate toward stories that emotionally affirm us: using empathy to demonstrate that we were right all along.

To see this in action, try out the *Wall Street Journal*'s "Blue Feed, Red Feed." This project aggregates left- and right-leaning Facebook posts. Readers can choose a political issue and see what people on either side find when they search for it on the social media platform. The results could be a dictionary illustration of confirmation bias:

Facts and statistics vary wildly across red and blue feeds. Even more than that, each feed drives readers' emotions in opposite directions. Choose abortion, and *Blue Feed, Red Feed* shows you two worlds of victims. On the left, women's rights are under siege. A woman in El Salvador was sentenced to thirty years in prison after her child was stillborn, because of that country's total abortion ban. On the right, babies' body parts are harvested by doctors. Or try immigration: On the left, children are ripped from their families. On the right, a murderer who immigrated here illegally is acquitted.

The effect is strangely symmetrical. A reader feels sickened, saddened, and outraged—all based on their empathy for victims—but each side's victims are the other's perpetrators. Of course, we don't need the Internet to create this effect. In the Jim Crow South, lynch mobs often began with pity for a white woman allegedly (though often not actually) raped by a black man. But online environments feed hatred like accelerants on a flame. In one recent survey, people reported that they felt more outrage when skimming the Internet than when engaged with newspapers, television, or in-person conversation.

We are also *seen* differently online, especially when we replace our names and faces with user names and avatars. Anonymity has its benefits: It allows people to safely organize protests in totalitarian nations, and to discuss their sexual identity without fear of being outed. But it also strips away a key pillar of kindness. As we saw earlier, when people are accountable to one another—for instance, in small communities—cruelty becomes socially expensive. Anonymity frees people from these constraints, cutting the brake lines on social exchange. The Internet is filled with the resulting wrecks.

People who shot paintballs at Wafaa Bilal and those who wish each other violent deaths in comments sections work under cover of virtual darkness. Trolls spend vast amounts of time and energy sowing pain. Anonymity tempts people to try on cruelty like a mask, knowing it won't cost them. It does, of course, cost their targets. Online harassment can follow people into their homes, their rooms, and their beds. This might help explain why teenagers who are cyberbullied contemplate and attempt suicide even more than victims of traditional bullying.

When people do put their names and faces online, the person they present to the world can be quite different from their analog selves. Social media tempts us to present airbrushed versions of our lives. Spending time on Facebook tends to leave people more depressed, and this is likely one reason why: In their posts, our acquaintances and ex-colleagues eternally whitewater raft at sunset, while we sit at a desk under fluorescent lights, supposedly working but actually looking at them.

Social media also encourages us to broadcast fury toward outsiders. When people tweet emotionally about political and moral issues, their messages are shared more often—especially among people who already agree with them. Retweets are Twitter's most valued prizes: vague, tiny doses of approval that reinforce whatever produced them. This makes Twitter catnip for human tribalism. Old media gave us loud, extreme voices, and we gravitated toward them. New media encourages us to *be* those voices.

NONE OF THIS happens by accident. When empathy disintegrates online, it's often because someone designed it to. In 2016, fake news campaigns promulgated rumors in the run-up to the U.S. presidential election. Russian trolls didn't merely present misinformation; they went straight for America's racial, religious, and economic fault lines. Ads targeted people's tribal identities, and sowed fear and anger toward the other side—abetted by Facebook and other platforms. Many people viewed this planted content as more trustworthy than traditional media. Informational democracy disrupted national democracy, using our emotions as a vehicle.

Facebook and Twitter pay shareholders not by making users happy, but by keeping them online. They do so using artificial intelligence algorithms, which have access to a dizzying amount of information about our lives online—the sites we visit, what we post, the questions we ask Google in the middle of the night. They use it to keep us scrolling. Oftentimes, this means appealing to our frailties. A bipolar individual on the verge of a manic episode might be flooded with ads for Las Vegas vacations. A Facebook skirmish between family members might attract eyeballs, the same way people rubberneck at a highway crash. If it does, Facebook's algorithms will duly place similar exchanges at the top of our feeds next time we log on.

In 2017, Chamath Palihapitiya, an early Facebook executive, expressed "tremendous guilt" about the technology he helped create. "I think we have created tools that are ripping apart the social fabric," he told an audience at Stanford. "The short-term, dopamine-driven feedback loops we've created are destroying how society works." And yet, the same year, a start-up called Dopamine Labs (now renamed Boundless Mind) received a million dollars in seed funding. Founded by two neuroscientists, it uses principles of reward and reinforcement to keep people online. Its promise to clients: "Dopamine makes your app addictive."

As Asimov predicted, we've withdrawn from nature into the arms of "an environment that suits us better." Sadly, life online suits our worst instincts. It's tempting to blame technology itself for its effects on us, and many people do. Its major currents have certainly driven us further apart. But technology does only what its creators—and ultimately its consumers—ask of it. Some enterprising individuals are battling its empathy-reducing effects, using technology to create connections that would have been hard to imagine even a decade ago.

YOU WAKE UP on a bus, surrounded by all your remaining possessions. A few fellow passengers slump on pale blue seats around you, their heads resting against the windows. You turn and see a father holding his son. Almost everyone is asleep. But one man, with a salt-and-pepper beard and khaki vest, stands near the back of the bus, staring at you. You feel uneasy and glance at the driver, wondering if he would help you if you needed it. When you turn back around, the bearded man has moved toward you and is now just a few feet away. You jolt, fearing for your safety, but then remind yourself there's nothing to worry about. You take off the Oculus helmet and find yourself back in the real world, in Jeremy Bailenson's Virtual Human Interaction Lab at Stanford.

For more and more people in Silicon Valley, a long and dangerous bus ride isn't a simulation; it's reality. Santa Clara County—home to Facebook and Google—contains the nation's second highest concentration of affluence. The soaring cost of living here has displaced all but the wealthiest. In Palo Alto, the nation's tech epicenter, the number of homeless people has increased by a staggering 26 percent in the past two years, with higher concentrations of children and families among them. They turn to shelters, campers, and, in harder times, bus line 22.

Just a mile from Stanford's bucolic campus, the 22 departs Palo Alto for San Jose, and shuttles between the two cities all night. Silicon Valley's homeless have taken to it for safety and shelter so often and in such numbers that it's been dubbed Hotel 22. Dozens of people shuffle on past midnight, in an orderly, exhausted procession. They take the ninety-minute ride from one end of its route to the other, get off, and then get right back on. Drivers on line 22 know the drill. After leaving the first station, one announces over the bus's intercom, "No lying down, no putting your feet on the seats. . . . Be respectful to the next people getting on because they're going to work. Let's have a nice safe ride; let's do it right. Anybody wants to act up, well, you know the consequences."

On Palo Alto's El Camino Real, 2019 Tesla coupes sit in a dealership, waiting to be scooped up by new multimillionaires. Down the road, rows of RVs are parked for days on end, housing families whose lives were upended weeks or months ago. When we're no longer shocked by juxtapositions like these, it means we have gone blind to the suffering of homeless people. Sometimes we fail to see their humanity at all. In one study, neuroscientists showed people images of individuals from many groups—businesspeople, athletes, parents—while scanning their brains with fMRI. Parts of the brain associated with empathy were activated by every group *except* the homeless.

Acknowledging the experience of homeless individuals is painful; it induces guilt; it damages our sense that the world is just. Circumstances like these tip the balance in empathy's tug-of-war, favoring avoidance. Jeremy and I set out to investigate whether we could use immersive technology to make caring about forgotten people easier, more natural, or even inescapable.

A few decades ago, the technology in Jeremy's lab existed only in science fiction. A few years ago, it was exclusive, expensive, and glitchy, an exciting idea with little useful execution. Then it exploded. In 2014, Facebook acquired Oculus VR for about $2 billion. At the same time a raft of cheap, portable devices, ranging from $10 to $300, made virtual reality accessible to the average person. According to Jeremy, this is no incremental shift in the media landscape. "VR is far more psychologically powerful than any medium ever invented," he writes. Its secret sauce is what Jeremy calls "psychological presence." Books and movies transport us into their stories, but readers and viewers remain aware that they are reading and viewing. VR envelops people so completely that they forget it is media at all. When people are immersed in VR, their hearts race as they fly over a city, they jump to avoid falling debris or enemy fire. They confuse virtual experiences for real ones, which makes sense, because those experiences are quite real to them.

VR enhances fantasy and will almost certainly define the future of gaming and pornography. But psychological presence can also allow us to try on *real* experiences. According to Jeremy, this is where the technology's true power lies. Quarterbacks use it to better visualize the field, and medical students use it to practice complicated procedures. In both cases, VR facilitates quick, deep learning. VR also allows people to see themselves in the body of an elderly adult or person of another race, or through the eyes of a colorblind person. As Jeremy and his colleagues have found, these experiences decrease stereotyping and discrimination.

Findings like these led the artist Chris Milk to celebrate VR as "the ultimate empathy machine." In 2014, Milk created *Clouds Over Sidra*, a VR film that tells the story of a twelve-year-old girl in the Za'atari camp in Jordan, home to about eighty-four thousand Syrian refugees at the time. Viewers "meet" Sidra and spend time with her and her family, exploring the camp along the way. Milk recently brought the film—and the Oculus headsets necessary to watch it—to the Davos World Economic Forum in Switzerland.

"These are people," he reflects, "who might not otherwise be sitting in a tent in a refugee camp . . . but one afternoon in Switzerland, they all found themselves there." According to Milk, "being there" mattered. He described why in a recent talk: "You're not watching it through a television screen. . . . You're sitting there with her. When you look down, you're sitting on the same ground that she's sitting on. And because of that, you feel her humanity in a deeper way. You empathize with her in a deeper way."

The idea is simple and powerful. Yes, technology can make it harder for us to see one another. But used differently, it can do just the opposite.

Milk's video makes for a powerful story, but few experiments have examined whether immersive technology *actually* builds empathy, and there are reasons to doubt. Imagine telling someone they have the chance to spend an hour inside the world of a refugee. Who would agree, and who would avoid it? Chances are, people who don't want to empathize wouldn't want to enter an "empathy machine" at all. VR might make already caring people care a tiny bit more. The question is whether it can do better than that.

About three years ago, Jeremy and I, along with our students Fernanda Herrera and Erika Weisz, decided to find out. We designed a VR experience to help Bay Area residents see their homeless neighbors in a new light. Using an Oculus Rift, viewers explored scenes that told the story of one person's descent into homelessness. A viewer first "wakes up" in his apartment, facing eviction, and is asked to take inventory of furniture he can sell to keep afloat. That fails, and the viewer finds himself living in his car. A police officer catches him staying there illegally and impounds the vehicle, and the viewer ends up on Hotel 22. In this last scene, he can also learn about his fellow passengers. If he "clicks" on the father and son next to him, a narrator explains, "This is a father, Ray, and his son, named Ethan. Ethan's mother suffered from a chronic illness and recently passed away. Left with the hospital bills, Ray is in debt. They're on a family shelter waiting list. So until free spaces become available, they sleep on the bus at night."

Jeremy and I were confident that people would feel empathy after walking around in a homeless person's virtual life. But would VR build *more* empathy than traditional approaches? To test this, we assigned some people to complete our VR exercise, and others to read the same story—eviction, impoundment, Hotel 22—while imagining what the protagonist would think and feel. This kind of perspective-taking exercise has successfully increased empathy in dozens of studies, meaning that in ours, VR had real competition. When we began the project, I bet Jeremy that VR wouldn't make people feel any more empathy than a low-tech alternative.

I was wrong. At first, both exercises increased people's empathy for the homeless, and even their willingness to donate money to local shelters. But when we tested their caring more strenuously, differences emerged. We told participants about Proposition A, a ballot measure that would expand the Bay Area's affordable housing initiative and also slightly increase taxes. People across our experiment said they supported the measure, but when we offered them the chance to sign a petition in support of it, those who had completed VR were more likely to agree. The technology also created longer-lasting empathy. A month after taking part in our study, participants who had undergone VR remained supportive of ballot initiatives to support the homeless and were less likely than other participants to dehumanize them.

Neither Jeremy nor I believe that VR is the perfect empathy machine. Some experiences simply cannot be mimicked. We can put someone on Hotel 22 for a few minutes, but we can't make them feel the grinding desperation of long-term hunger. Nonetheless, we're optimistic that VR can raise people's curiosity—driving them to learn more about people they'd otherwise ignore. Jeremy and his team have installed our Hotel 22 experience in malls and museums around the Bay Area, where thousands of people have tried it. In *To Kill a Mockingbird,* Atticus Finch advises Scout, "You never really understand a person until you consider things from his point of view . . . until you climb into his skin and walk around in it." As VR becomes more commonplace, millions of people will have the chance to do just that.

WHERE VR MAKES other people's lives more visible, other new technologies allow us to literally see their emotions. In 2012, Google rolled out its Glass project to intense hype and a quick demise. The idea sounded revolutionary: Through a small, transparent computer affixed to a pair of glasses, people could "augment reality," overlaying digital data onto the analog world. The possibilities seemed endless. As Glass users looked at a row of restaurants, each one's Yelp rating might float transparently in front of it. Train stations could be digitally tagged with the time of their next departures. Glass promised to make life more like a video game.

But when beta testers started wearing it, mostly in New York and San Francisco, they found themselves shunned. The tech looked awkward. It smacked of smug, early adopter superiority. Most of all, it made others deeply uncomfortable. It's creepy to think that the person next to you at a bar can snap pictures of you by winking. Weeks after Glass hit test markets, dozens of bars and restaurants banned its users, colloquially known as "Glassholes." As a mass-market product, it was dead on arrival.

Glass disappeared, but augmented reality did not. In one case, it has inspired a revolutionary tool to help people like Thomas Coburn. Thomas is a precocious, scruffy ten-year-old boy. He loves video games and cartoons—when I met him in December 2017, his T-shirt featured Snoopy in a Santa costume—and he has autism. He is intensely shy and avoids eye contact. He repeats movements over and over again. Years ago, he would lick his lips so much that they would bleed.

Thomas was diagnosed when he was eight, much later than most kids, in part because his challenges are so specific. He has no problems speaking, reading, or writing. He is not unemotional—in fact, he can become overwhelmed by his feelings, leading to outbursts and tantrums. Thomas's mother, Heather, says he "wears his emotions on his sleeve." Despite this expressiveness, he does have one feature typical of autism: trouble picking out emotions in other people's facial expressions.

Heather was saddened by Thomas's diagnosis, but more than anything she felt lost. "There was a disconnect, like, 'What do we do now?'" For decades, people didn't think kids with autism could do much of anything. According to the Roddenberry hypothesis, empathy is an immutable trait, such that people who have trouble empathizing will never be able to improve. Hans Asperger, one of the psychiatrists who "discovered" autism, believed as much, writing that children who had it would "surely suffer disconnection for the duration of their lives."

This fixed perspective on human nature is just as wrong for those with autism as for everyone else. Over the past thirty years, new therapies have helped kids like Thomas overcome their difficulties. By far the most popular technique, Applied Behavior Analysis (ABA), trains kids by breaking down behaviors into bite-sized pieces. Brushing one's teeth means first applying toothpaste. Paying attention to someone means facing them, standing still, and not speaking out of turn. ABA therapists practice each step with autistic people over and over again, rewarding them for getting each one right.

Proponents liken ABA to skill drills. The social world comes naturally to people without autism, the thinking goes, but autistic people must train for it like a football game or a violin recital. ABA's principles go back to B. F. Skinner, a renowned and controversial psychologist who founded the field of "radical behaviorism." Skinner used positive reinforcement to train pigeons to do everything from turning in circles to steering missiles via satellite images.

Critics of ABA think it's a crude method, better used on animals than people. But it produces remarkable results, even "curing" some children of all visible autistic symptoms. It also takes vast amounts of time—typically ten to forty hours a week—and at typical rates of $50 to $120 per hour, it is well outside the reach of most families. Private insurers and Medicaid now cover ABA in most states, but this doesn't guarantee access. Since the turn of the century, the number of kids diagnosed with autism has more than doubled, to one in every sixty-eight children. The number of ABA therapists has not kept pace, and poorer families often wait years for treatment.

Other autism therapies are less intense. Software packages such as Mind Reading gamify social interaction. Users browse faces, voices, and stories associated with different emotions. They quiz themselves, trying to tell emotions apart or spot them in new faces. Mind Reading and similar programs require only about two hours a week, and do improve people's ability to decipher feelings. But real-world empathy is messier and noisier than a video game, and the programs often fail to help people with autism navigate actual social interactions at school or home.

For empathy training to better help those with autism, it will need to look more like everyday life or, better yet, *become* part of it. Engineers at Stanford are betting that augmented reality can make that happen. This idea began with a tech prodigy who could have been cast in *The Social Network*. Catalin Voss grew up in Germany as a self-described "Apple fanboy." He saved his money to buy an early iPod Touch and began tinkering with its programming language. His simple iPod games evolved into a podcast that gave fellow programmers tips on creating apps. It quickly rose to number one on the iTunes store. Catalin was thirteen, but, as he puts it, "on the Internet no one knows how old you are." He received several job offers per week, and by sixteen left home for Silicon Valley as an emancipated minor. He enrolled at Stanford and hired his math teacher—Nick Haber—to work in his first start-up.

Together, the two built an education app to help gauge students' level of interest during classes. They used a technology called "affective computing," artificial intelligence algorithms that make educated guesses about how people feel based on their facial expressions or voices, getting it right 75 to 90 percent of the time. Affective computing is growing rapidly, reflecting its thrilling, sometimes frightening potential. Advertisers use it to test whether commercials and movie trailers gin up the desired feelings among consumers. In five years, your phone might monitor your mood the way a Fitbit monitors your heart rate, flagging signs of anxiety or depression. Apple recently acquired the affective computing start-up Emotient, fueling speculation that it hopes to make Siri more "human."

Catalin and Nick soon realized that their face-tracking technology was more than an educational tool. "We were building software to try to understand faces," Catalin recalls, "and I knew somebody . . . who really struggled with understanding social cues we take for granted." That was his cousin David, who has autism. Catalin realized that by combining his face tracker with Google

Glass, he could put affective computing to work for people like David, giving them a "cheat code" to what others around them were feeling. Silicon Valley investors were dismissive—autism was too small a market, they argued—but Dennis Wall, a professor of pediatrics and biomedical data science at Stanford's Medical School, took a chance on the project.

Catalin and Nick built a prototype, and Dennis's son tried it out. It was clunky and tended to overheat, but it was still a hit. Whenever Dennis's son looked at a face, a green box in the upper right corner of his Glass would appear. When the person in front of him smiled, frowned, or expressed any emotion, the Glass would calculate what that person most likely felt, and project it as an emoji in real-time—a red, angry face one moment, a yellow, frightened one the next. The effect is surreal: one part *Terminator*-style cyborg infographic, one part cartoon emotion à la *Inside Out*.

Dennis's son does not have autism; for him, the Glass emojis were entertaining but redundant reminders of emotions he'd already figured out. But the technology promised kids with autism much more. In 2016, the Autism Glass project rolled out its first home trials. Just days into their first round, Catalin received a note from one of their participant's mothers. Her son was making more eye contact when using the Glass, a change his teacher picked up on, too. "It's almost like a switch was turned," she wrote. "Thank you!! My son is looking into my face."

The Glass team worked closely with families, tinkering with the platform in response to their feedback. The project's first trial included Thomas and his family along with twenty-four others. Thomas wore the Glass for about twenty minutes, three times a week, throughout the summer of 2017. Almost immediately, it sharpened his understanding of family members' feelings. Thomas sometimes does only the bare minimum on chores, angering Heather. After doing so while wearing Glass, he glanced up at her. She remembers his response: "He was like, 'Oh. Mom's *upset*!'"

Even months after Thomas finished using Autism Glass, Heather finds herself yelling less, because he picks up on her frustration more quickly. A couple of weeks into the trial, his teacher also noticed that Thomas was more attuned to his classmates and was dealing more effectively with conflicts. The Stanford team's preliminary results suggest that after using Autism Glass, kids get better at understanding others people's emotions. The Glass team is now replicating this finding with a larger trial of fifty families.

Autism Glass's algorithms don't always work. Thomas's father has a thick beard, and Glass often fails to pick up his expressions. But the project has given Thomas's family more reason to talk about feelings and where they come from. It has also taught Thomas's parents about how they look from their son's perspective. Imagine having a camera trained at you while you do dishes or try to get your child to make his bed: You might realize that you look more irritated than you mean to. This is what happened to Heather and her husband. "We're more aware of our RBF [resting bitch face]," she says, "and how we're portraying ourselves to other people."

The Autism Glass team has created new features that encourage autistic kids and their families to discuss emotion. In their parent-review app, families can look through video the Glass recorded, color coded to identify heated moments. They can talk about what happened right before someone became upset or happy or bored, and use that knowledge in future interactions.

Catalin and his team hope that evidence from their clinical trial will allow Glass to become a reimbursable, insurance-covered treatment device for autism. Google Glass, still futuristic and alien, remains unpopular. I've worked in the heart of Silicon Valley for six years and have yet to see anyone wearing Google Glass outside of this project. Still, for thousands of children who lack the autism services they need, this technology could become far *more* accessible—and cheaper—than traditional therapies such as ABA.

If it does, it will bring with it a new vision of treatment for autism. Autism Glass does more than train a child in how to recognize emotions. It teaches parents, siblings, friends, and caregivers how to manage them together.

> I'm working on a big project and am scared I won't be able to finish on time. I keep getting distracted and then become angry at myself for not making progress, sometimes to the point of panicking. Also, finishing it on time is important to supporting my family, so I feel like I'm letting them down.

I typed this confession into Facebook's Messenger app as the deadline for the book you're reading careened toward me. My progress felt glacial, and I was getting scared. Most of my friends and colleagues had no idea. When someone asked, "How's that book coming?" I'd project confidence—in part not to complain, in part to try it on. Now I was opening up, not to an old pal, but to Koko, a bot that helps strangers help one another.

"OK, thanks for sharing all that," the algorithm replied. *"Sending this off to the Koko community. . . . You should start getting replies in a few minutes."* Really? Were there strangers somewhere in the ether standing by to read about my anxiety? Would they say anything useful? Shouldn't I be writing my book instead of whining online?

"While you wait," Koko continued, *"let's try helping others."* Now it made more sense; I *was* one of the operators standing by. I agreed, and Koko walked me through a crash course in empathic listening. Don't tell people what to do; do respect their feelings. Try to help them see the bright side. Conscripted into Koko's helper army, I read notes from break-up survivors and bullying victims. The writers seemed younger than me—most Koko users are in their teens or early twenties—and reminded me of times in my life when social rejection felt like a death sentence. I did my best to be encouraging; I said that things do get better. It felt wonderful, the opposite of wasted time.

"Someone replied to your post!" Koko cheerily announced. Six minutes had passed since I'd first vented.

> First, you're working on a big project! That's awesome and brave. Big projects are hard but rewarding. So expect all these things— getting distracted, getting angry. Your concerns show you ARE a very considerate person to your family. Someone who's not considerate simply wouldnt [*sic*] care. One last thing that helps for me . . . sometimes our brains need to get distracted a little, it's their way of telling us to pause and breathe.

I was moved. A stranger maybe half my age had typed this insightful, kind missive. We were anonymous to each other and would stay that way; Koko users have no way of connecting beyond the messages they send. That made our dialogue feel *more* intimate, like a virtual cubbyhole. Koko is a vast constellation of connections just like this, and a different vision of what social networking can be. It offers a place where we can be vulnerable rather than boastful, supportive rather than vengeful—where we can be seen in new ways.

Koko was conceived when its creator decided to depend on the kindness of strangers. In 2012, Rob Morris was a PhD student at MIT's Affective Computing Group, which specialized in tools very much like those Catalin Voss and the Glass team use. He dabbled in several projects—could he help machines detect stress and anxiety, or predict which types of music would evoke strong emotions in listeners?—but didn't make much headway. "I was swimming in circles," he remembers.

Rob had long struggled with depression; in Cambridge, it gripped him more tightly. The city was dark and cold, and he found MIT severe, right down to its architecture ("It's like every building terminates sharply, in a scalpel"). He had severe migraines, and the medication he took in response left him groggy and dull. He also felt intellectually outmatched. "The people sitting next to me

[were] some of the best engineers in the world, and I [was] sitting there trying to figure out how to write a for-loop [a basic programming command]." He didn't wonder *if* he'd have to drop out, only when.

On a whim, Rob joined a group of computer scientists interested in "crowdsourcing." In 2005, Amazon launched Mechanical Turk (MTurk for short), an online marketplace where "requestors" post simple jobs known as Human Intelligence Tasks (HITs), and "workers" complete those tasks in exchange for pay. HITs tend to be mundane. A lot of them look like the CAPTCHAs people must complete before entering a website, asking workers to identify numbers, letters, and objects in pictures. Scientists use this information to train artificial intelligence algorithms to see more like people do—using human workers with the goal of making them obsolete.

At MIT, scientists thought about HITs more creatively, as a sort of virtual prosthesis made up of human mental energy. One of Rob's friends created a Microsoft Word plug-in that crowdsourced editing. Writers who needed to cut the fat from a document could say how much shorter they needed it to be. Their text would be sent to workers, who would find extraneous words and suggest succinct rewrites. Requestors could watch while the document was shortened in real time. Another app allowed visually impaired people to take pictures of text, which would be sent to MTurk, transcribed by workers, and read aloud in seconds. On the surface, these apps look like magic, but they contain a crowd's worth of human computation under the hood.

This was Rob's eureka moment. As an undergrad at Princeton, he had studied with Daniel Kahneman, a Nobel laureate who has spent his career exploring human irrationality. Kahneman taught Rob that people's thoughts and feelings often contain distortions and errors—in computer science terms, bugs. Rob knew his depression reflected similar bugs. After failing at a task, he'd feel worthless, as though he would never accomplish anything. Distortions like this are common in mental illness. They're usually treated with cognitive behavioral therapy (CBT). CBT therapists help people rethink their problems and realize when they're blowing them out of proportion. Online platforms also allow people to do CBT exercises on their own.

Rob had tried both, but with limited success. "The hardest thing about CBT," he remembers "[is] trying to wrestle with your own thoughts at the moment when your thinking is most impaired." It was like trying to use broken code to debug itself. While learning about crowdsourcing, Rob realized he didn't need to work against his sadness and fear alone. "Holy shit," he thought. "I can send my [thoughts] out, and get dozens of people to work on them."

The Internet has long been a haven for people who suffer in ways their friends and neighbors might not understand. Consider the paradox of rare illnesses, such as cystic fibrosis or myasthenia gravis. Each occurs in less than one out of every thousand people, but there are hundreds of such illnesses. That means that more than one in every ten people suffers from health concerns that no one around them shares. Millions of them have turned to online communities, including Facebook groups and message boards such as RareConnect.org. Patients share tips on managing symptoms, dealing with insurers, and exploring new therapies. But more than that, online illness communities are wells of mutual empathy and understanding. Rare illness patients who feel isolated, judged, or just "different" find solace in people they will never meet in person.

Almost anyone with unusual preferences, hobbies, or experiences—who years ago might have felt "weird"—can find a community online. NASCAR or knitting aficionados, chronic fatigue sufferers, and trans youth all support their own. Their empathy is powerful but idiosyncratic, based on shared experience.

Seeing what his colleagues were doing with MTurk, Rob realized he could broaden Internet-based kindness beyond rare communities so that anyone could benefit. He cobbled together an unusual HIT. Workers would first learn about mental bugs common to depression. They'd then read Rob's inner monologue, trying to pick out bugs and reframe his situation in a sunnier light. Late one night, Rob wrote down his anxieties like he had many times before: He was a moron; he didn't belong at MIT; he'd soon drop out, penniless. But this time, he hit SEND, transmitting them into space. He refreshed the results page every few seconds, not knowing what to expect.

After a few minutes, responses began pouring in, more thoughtful than Rob could have hoped for. Workers gave him new ways of thinking about his feelings—optimistic, but also practical. "It was so beautiful that some stranger would be able to open a door, enter this very private place, and kind of clean things up for me." He was moved to tears. Something else happened that night, too. At the end of HITs, workers typically leave comments, usually pointing out technical issues or complaining that the job was boring or underpaid. But after engaging with Rob's problems, workers were energized. They thanked him for the opportunity and said they would do more HITs like that for free.

Rob had thought of his HIT as a private tool, where he could trade money for help with his thinking. Perhaps, he thought, he'd share it with a few friends. "They could put coins in the machine, and out would come these beautiful responses that would hopefully be uplifting." Instead, he had struck a deep well, filled with not only collective intelligence but also collective goodwill.

Today, an Instagram search for the hashtag #depression yields more than thirteen million results; #suicide produces nearly seven million. As Rob puts it, at any moment you can find a "never-ending stream of misery" on this and other social media platforms. Countless people, many of them young, confused, and isolated, would benefit from CBT but can't afford it or simply don't know about it. Many of them feel comfortable disclosing their pain only online, to strangers. The same anonymity that gives trolls cover to be cruel gives many others permission to be vulnerable.

Rob realized that people all over the world needed help, and people all over the world wanted to help others, and he could connect them. "It was electric," he remembers. After a "couple of years of horrible agony" building the platform, it was ready, and Rob began testing it out. He recruited helpers and trained them—as he had before—in basic CBT. Afterward, they practiced writing to people in need. A separate set of workers rated how empathic their responses seemed. Just minutes of training sharpened helpers' empathy.

Next, Rob opened the platform up to people suffering from depression. They wrote about their feelings for about twenty-five minutes a week for six weeks. Then they either shared them with helpers or kept their writing private. Both groups experienced less depression by the end—expressive writing has long been known to improve mental health—but the people who interacted with helpers demonstrated a stronger ability to rethink their problems.

Helpers also benefited from engaging with other people's problems. Rob noticed this with the very first person who logged on, a severely depressed young woman. Her first post was not only grim but also poorly written and disorganized, as if her mind were engulfed in fog. Then her turn came to help someone else. Suddenly, her writing became incisive and elegant. "It was like she was a different person."

We tend to think of kindness as benefiting others at a cost to ourselves. Helpers agree to suffer so that others can hurt less, like the staff at UCSF's ICN or empathic parents whose health dwindles as they care for their children. Other times, however, helpers benefit by giving to others. Generosity leaves givers fulfilled and less stressed, and it even lengthens the lives of elderly volunteers. My colleagues and I have found that this is especially true when givers experience empathy for the targets of their goodwill.

Rob's platform offers people another way to benefit from their own empathy. A breakup or negative performance review feels like the end of the world when it happens to you. When it happens to someone else, it's easier to imagine things getting better. That's exactly what his helpers found. By rethinking one another's experiences, they became sharper empathizers. Later, if they encountered issues themselves, they were able to turn their new, more hopeful lenses back on themselves.

In 2011, Rob teamed up with Fraser Kelton and Kareem Kouddous to grow his academic project into Koko. "Kokobots" now inhabit about a dozen large social networks, including Facebook and Twitter, and Kik, an instant messaging app popular among teenagers. About a million people have used the program. Not everyone who signs up for Koko has good intentions; 10 to 20 percent of responses are unhelpful or abusive. But these do no harm. Where typical social networks have public feeds, avatars, and flame wars, Koko is a massive web of conversations, each comprising just two beats: a disclosure and an attempt to help. Rob and the Koko team use a combination of artificial and human intelligence to identify negative content and pluck it out of conversations in real time. Trolls can spit venom all they want, but unbeknownst to them it evaporates in the space between people, never reaching its target.

To the extent that social networks count on raw traffic, cruelty is a reliable fuel. Koko capitalizes on the idea that supportive, private interactions might be as addictive as angry, public ones. Anonymity can make kindness hard, but Koko uses it to pave the way for genuine empathy between strangers.

A lot hinges on whether other tech companies follow suit. Mark Zuckerberg famously instructed his employees to "move fast and break things." By now it seems clear that they have broken quite a lot. Technology has ripped us apart, but it can also create new opportunities for us to come together. The way we decide to use it will determine the fate of empathy for decades to come.

The Future of Empathy

IN JUNE 1944, General George Patton was given command of the Third Army, a group of mostly inexperienced troops who would soon be fighting in Nazi-occupied France. In a series of legendary speeches, Patton tried to whip his men up until they forgot their fear, or at least pretended to. Wearing a gleaming helmet, with an ivory-handled .357 Magnum at his hip, he talked about valor, duty, teamwork, hatred of the enemy—whatever would inspire them to fight and to win. He was bombastic, and so crass that the 1970 biopic *Patton* had to heavily whitewash his language. But at the end of a tirade on the killing and dying to come in the weeks ahead, Patton exhorted his troops to think further:

> Thirty years from now when you're sitting by your fireside with your grandson on your knee and he asks, "What did you do in the great World War Two?" You won't have to cough and say, "Well, your granddaddy shoveled shit in Louisiana." No, sir, you can look him straight in the eye and say, "Son, your granddaddy rode with the great Third Army and a son-of-a-goddamned-bitch named George Patton!"

In this book, we've toured empathy's battlefields. We've seen the forces that push us toward hatred and indifference, and seen people push back against them. Many of them have won, defeating their own estrangement, toxic cultures—even actual war—to reclaim their humanity and discover each other's. But the war we're fighting is much bigger. Our empathy is the legacy we leave generations to come, who must live in the world we leave behind.

How can we be the ancestors they deserve? This question drives Ari Wallach. As a consultant for businesses, governments, and nongovernmental organizations, Ari began to notice that his clients were focusing on shorter and shorter time horizons. Groups that had once asked him for help with their two-decade plan now wanted to talk about just the next six months.

Ari calls this "single marshmallow thinking," after the iconic research of Walter Mischel, in which children had trouble waiting for two marshmallows rather than impulsively consuming one. Single marshmallow thinking doesn't stop after childhood. Adults find it difficult to opt for kale over burgers or retirement savings over credit card spending; companies focus on quarterly earnings to the detriment of their long-term outlook.

Short-term thinking is not always irrational. Children who can't trust in promises of future marshmallows are wise to gobble up one now instead of waiting. If we tear each other apart in the next twenty years, there's no use in planning for the next two thousand. But many of humanity's existential threats—climate change, water shortage, overpopulation—are building more slowly. They might not personally affect us, but they will shape the lives of our grandchildren, and theirs, and theirs. We're doing a woefully inadequate job of protecting future generations, in part because it's so difficult to imagine them.

In this book, I've expressed the hope that people might widen their circle of care to encompass all of humanity. But preserving our future requires expanding it even further, not just across space but across time. This is a core mission of a new movement known as effective altruism (EA). Effective altruists turn morality into math, by calculating how each of us can make the greatest possible positive impact on the world. They think about the future a *lot*. The philosopher Nick Bostrom writes that, barring disaster, the earth should sustain life for another billion years. If he's right, this planet will be home to ten million billion people over that span. Future humans will outnumber the people living on earth today by a million to one. According to effective altruists, we should care vastly more about them than we do about ourselves.

It's hard enough to empathize with people who are alive right now but who are distant or different from us. How can we be expected to care about hypothetical people we will never know? Peter Singer suggests that we leave our emotions out of the equation. Effective altruists, he writes, "don't give to whatever cause tugs strongest at their heartstrings. They give to the cause that will do the most good."

Paul Bloom, the author of *Against Empathy*, is more forceful. He argues that empathy *prevents* us from caring about the future, because it's hardwired for the here and now. On climate change, Bloom writes: "Empathy favors doing nothing. If you do act, many identifiable victims—real people whom we can feel empathy for—will be harmed by increased gas prices, business closures, increased taxes, and so on. The millions or billions of people who at some unspecified future date will suffer the consequences of our current inaction are, by contrast, pale statistical abstractions."

But as we've seen throughout this book, we are capable of more. We can *choose* empathy, making our descendants feel more real and their well-being more urgent.

That's the bet Ari is making. In 2013, he founded Longpath Labs, a group devoted to cultivating sustainable thinking. He works with psychologists, myself included, to encourage intergenerational empathy. Some of his techniques are similar to ones we've encountered in this book. In a recent Longpath workshop, people were asked to introduce their great-great-grandchild. They first spent time writing about their descendant's life. "What is their name?" the worksheet reads. "Occupation? What character traits and purpose define them? What do they struggle with?"

Ari also asks people to consider the past and the sacrifices our ancestors made for us. The psychologist David DeSteno has shown that gratitude protects against short-term thinking. In one study, he and his colleagues asked people to recall an event that made them feel grateful or one that made them feel happy. Participants then made choices between taking small amounts of money now versus larger amounts later, in a grown-up version of the marshmallow test. Those who recalled feeling grateful made more prudent decisions.

This effect can spread across generations. In another study, people read about a company that limited its fishing decades ago so that more resources would be available today. Afterward participants were more willing to sacrifice their own prosperity to benefit others in the future—a kind of golden rule across time. There are, of course, reasons not to feel grateful toward generations past, in their legacies of bigotry or debt. But there are also countless cases, both personal and large-scale, in which our elders scratched and bled to create a better world. If we can remind ourselves of those choices, we might be inspired to do the same for those who come after us.

Another feeling that cultivates long-term thinking is awe—the experience of something so vast that it interrupts our everyday preoccupations. In his *Pale Blue Dot,* Carl Sagan shows readers an image of the distant Earth, taken by Voyager 1 on Valentine's Day 1990. Three grainy streaks cut through the darkness. One of them is punctuated with a minuscule bright speck. In the passage that follows, Sagan writes:

Look again at that dot. That's here. That's home. That's us. On it everyone you love, everyone you know, everyone you ever heard of, every human being who ever was, lived out their lives. The aggregate of our joy and suffering, thousands of confident religions, ideologies, and economic doctrines, every hunter and forager, every hero and coward, every creator and destroyer of civilization, every king and peasant, every young couple in love, every mother and father, hopeful child, inventor and explorer, every teacher of morals, every corrupt politician, every "superstar," every "supreme leader," every saint and sinner in the history of our species lived there—on a mote of dust suspended in a sunbeam.

In the face of the universe and its vastness, our own concerns become vanishingly small. This is a fearsome realization, but it also combats selfishness. Psychologists induce awe in people by showing them enormous things—a towering grove of redwoods, a skyline of Himalayan peaks, the Milky Way, the sweeping vistas of *Planet Earth.* Afterward, people report feeling smaller, but also more connected to the rest of humanity, and they act more generously toward others.

There's no reason the vastness of time should be any less potent. In southern France's dolmen d'Er Roh, archeologists discovered Neolithic gold beads: tiny, with intricate striping and spiral patterns molded into them about four thousand years ago. Touching one of those beads, we might imagine the person who crafted it a hundred generations ago, and the person who might hold it a hundred generations from now. We are a link in an enormous chain of humanity; remembering this might incline us to tend to its future.

Of course, we won't be part of that future. Considering distant generations also means facing our own mortality: an unenviable experience that can take us to some dark places. Psychologists who study "terror management" force people to peer into the void—for instance, by writing about how it will feel to die. When people experience anxiety in response, they seek out whatever makes them feel safe, including the comfort of their own tribe. They show more interest in extremist propaganda and become more hostile to outsiders.

But conversations about mortality don't have to begin and end with fear, or even with death. Questions of the afterlife aside, we all live on in the legacies we leave; that thought alone can prompt people to help future generations. Consider Alfred Nobel. In his day, he was a prolific arms dealer who patented hundreds of explosives and invented dynamite. In 1888, Alfred's brother, Ludvig, died. As the story goes, a French newspaper, receiving an erroneous tip that it was Alfred who had perished, published a premature obituary. The title read "Le Marchand de la Mort Est Mort," or "The Merchant of Death Is Dead." Nobel was apparently so disturbed by this that he secretly bequeathed most of his fortune—about $250 million in today's dollars—to establish the Nobel Prize, and forever change his legacy.

Ari finds this a crucial part of building Longpath thinking. In his workshops, people write their own eulogies. "Try and push past your own life," he said in a recent talk, "because if you can, it makes you do things a little bit bigger than you thought were possible."

Empathy—in its ancient form—is built on self-preservation. We care for our children because they carry our genes, and for our tribe because they offer sustenance, sex, and safety. Building concern for a future that has forgotten us runs counter to our Darwinian impulses. But there are still ways to cultivate that concern. If we can, we'll be evolving our empathy in real time, into something larger and more enduring.

It's easy to live in a less intentional way. Building a new sort of empathy takes effort and sacrifice, for people who might not repay it. But in the face of escalating cruelty and isolation, we are fighting for our moral lives. Doing what's easy is seldom worthwhile, and in moments like these, it's dangerous. We each have a choice, and the sum of our choices will create the future.

What are you going to do?

ACKNOWLEDGMENTS

I'VE WANTED TO write a book since I was in high school. For all those years spent thinking about this process, I was remarkably wrong about how it would go. Most of all, I figured it would be solitary: a mental trek into the deep woods. Instead, it connected me with dozens of new people and deepened my ties to just as many old friends—all of whom made this book what it is.

The people featured in these pages gave hours or days of their time in interviews and allowed me to share their personal, often painful stories. I thank (in order of appearance) Ron Haviv, Ed Kashi, Tony McAleer, Emile Bruneau, Angela King, Sammy Rangel, Stephanie Holmes, Orrie and Ella of the Young Performers Theatre, Thalia Goldstein, Betsy Levy-Paluck, Bob Waxler, Bob Kane, Liz Rogers, Melissa Liebowitz, and the staff and families at the UCSF ICN, Albert Wu, Eve Ekman, Sue Rahr, Rex Caldwell, Joe Winters, and the staff at Washington's CJTC, Jason Okonofua, Catalin Voss, Nick Haber, Jena Daniels and Dennis Wall of the Autism Glass Project, Heather and Thomas Coburn, Rob Morris, and Ari Wallach.

This book also exposed me to countless things I didn't know I didn't know. Many researchers and scholars generously fielded emails and calls about all matter of arcane empathic marginalia. I'm indebted to Anthony Back, Simon Baron-Cohen, Daniel Batson, Daryl Cameron, David Caruso, Mark Davis, Lisa Feldman-Barrett, Adam Galinsky, Justin Gardner, Adam Grant, Daniel Grühn, Elaine Hatfield, Christian Keysers, Sara Konrath, Nour Kteily, Françoise Mathieu, Brent Roberts, Robert Sapolsky, Steve Silberman, Tania Singer, Linda Skitka, Sanjay Srivastava, Maya Tamir, Sophie Trawalter, Jennifer Veilleux, Johanna Voldhardht, and Robert Whitaker.

Several friends read draft sections of this book and provided indispensable feedback. Thanks especially to Lauren Atlas, Mina Cikara, James Gross, Yotam Heineberg, Ethan Kross, Adam Waytz, and Robb Willer. Max Thorn provided helpful fact-checking. And it was an immense pleasure working with Kari Leibowitz, who spearheaded the Evaluating the Evidence appendix and also provided deep and thoughtful commentary on other chapters.

Science is a team sport, and I've benefited immensely from working alongside brilliant and supportive collaborators throughout the years. Coauthors on research described in this book include Jeremy Bailenson, Niall Bolger, Carol Dweck, Valeria Gazzola, Fernanda Herrera, Matthew Jackson, Brian Knutson, Ihno Lee, Matt Lieberman, Jason Mitchell, Matthew Sacchet, Karina Schumman, Tor Wager, and Jochen Weber. My graduate mentor, Kevin Ochsner, took a chance on me in 2005, though I had little research experience, poor undergraduate grades, and a mess of half-baked ideas about empathy and the brain. His wisdom and friendship have made me the scientist I am now, and are all over these pages.

These days, I'm honored to advise passionate, generative young scientists in my own shop. I'd like to thank all past and present members of the Stanford Social Neuroscience Lab for their inspiration and hard work, and especially acknowledge Molly Arnn, Ryan Carlson, Rucha Makati, Sylvia Morelli, Erik Nook, Desmond Ong, Emma Templeton, Diana Tamir, and Erika Weisz, whose research is featured here. Erika deserves special recognition for the empathy-building interventions described in Chapters 2 and 6, which comprised the core of her dissertation.

Seth Fishman, my literary agent, and I have been friends for a decade. When I decided to wade into nonacademic writing years ago, he generously accepted the role of guide. He taught me how to pitch magazines; he read all the pieces I thought had turned out well and showed me how to *actually* make them work. He was an early, steady, and reassuring sounding board as I prepared the ideas behind this book.

Seth also connected me to an amazing team at Crown, for which I will always be grateful. Zachary Phillips provided quick, responsive support throughout the process. Meghan Houser conducted multiple close reads of the book at a late stage, providing incisive feedback that helped it across the finish line. She also encouraged me to write more about Star Trek TNG, which of course I'd never refuse. Many thanks, also, to Penny Simon, Molly Stern, Kathleen Quinlan, and Annsley Rosner for their invaluable support.

It's difficult to describe the impact that my editor, Amanda Cook, has had on this project and on me as a writer. She helped me find a deeper, more meaningful book hiding in my original ideas. She also exhibited saintly patience as I muddled through false starts, neuroticism flare-ups, overthinking, underthinking, and last-minute second-thinking. Firm, confident, but always compassionate, Amanda was the best partner I could have hoped for in this adventure. I cannot (and don't want to!) imagine this book in any other editor's hands.

Luke Kennedy and I met at a basement party in our junior year of college; since then he's the closest I've had to a brother. He's a steadfast pal and brilliant consigliere—ready to toss around any idea at any time, no matter how crazy it might sound. He's neither an academic nor a writer, but our hundreds of conversations have shaped virtually all of my thinking over the years.

My parents, Pervez and Iris, and my stepmother, Kathleen, have supported me through countless ups and downs, and we've been through some together. I am especially grateful that they supported my decision to write, here, about some of our most difficult times.

I began work on this book a week after my daughter Alma was born—in stolen, bleary moments while she slept. Alma and her sister, Luisa, don't know about the sacrifices they made for it, but they made them nonetheless. My wife, Landon, *does* know about the sacrifices she's made. In starting it while we had a newborn (and then another!), I subjected our family to more pressure at a time already stuffed with it. Landon spent immense amounts of energy to make space for me to finish it, all the while talking me through nagging doubts and full-blown freakouts.

For that and so much more, I dedicate this book to her. Landon is an inspiring psychologist, with thrilling projects on the horizon. Now it's her turn.

What Is Empathy?

You keep using that word. I do not think it
means what you think it means.

—INIGO MONTOYA, *THE PRINCESS BRIDE*

MOST OF US think we know what "empathy" is, yet we often mean different things when we use it. Psychologists have (sometimes heatedly) debated its definitions for decades. But despite our quibbling over details, most empathy researchers agree on the big picture. In particular, empathy is not really one thing at all. It's an umbrella term that describes multiple ways people respond to one another, including *sharing, thinking about,* and *caring about* others' feelings. These pieces, in turn, go by several names:

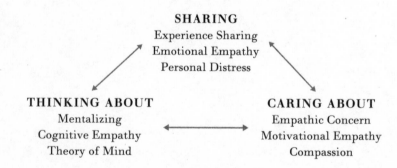

SHARING
Experience Sharing
Emotional Empathy
Personal Distress

THINKING ABOUT
Mentalizing
Cognitive Empathy
Theory of Mind

CARING ABOUT
Empathic Concern
Motivational Empathy
Compassion

Let's tackle these one at a time. To do so, imagine you're a senior in college, walking with a close friend to his apartment. On your way in he checks his mailbox, then freezes. "Holy shit," he says. "This is it." You know what he means. You've seen him work relentlessly for three years in hopes of getting into medical school, and into one program in particular. He's talked with you maybe thirty times since applying, alternately anxious, hopeful, or both. You rush upstairs, and he opens the envelope. His face contorts, and you lean forward, for a moment not knowing whether he's ecstatic or upset. It becomes apparent that he is not crying happy tears.

SHARING

As your friend collapses into a heap, you might frown, slump, and even tear up yourself. Your mood will probably plummet. This is what empathy researchers call *experience sharing*: vicariously taking on the emotions we observe in others. Experience sharing is widespread—people "catch" one another's facial expressions, bodily stress, and moods, both negative and positive. Our brains respond to each other's pain and pleasure as though we were experiencing those states ourselves.

Experience sharing is the closest we come to dissolving the boundary between self and other. It is empathy's leading edge. It is evolutionarily ancient, occurring in monkeys, mice, and even geese. It comes online early in life: Infants mimic each other's cries and take on their mothers' distress. And it occurs at lightning speed. Seeing your friend grimace, you might mimic his face in a fraction of a second, and parts of your brain associated with feeling pain might come online just as quickly.

Experience sharing also forms the foundation of empathy science. Even before the word "empathy" existed, philosophers such as Adam Smith described "sympathy," or "fellow feeling" in ways that tightly match experience sharing. Smith, for instance, writes that "by changing places in fancy with the sufferer . . . we come to either conceive or to be affected by what he feels." From "emotion contagion" in psychology to brain mirroring in neuroscience, experience sharing has long been the most famous piece of empathy.

THINKING

As you share your friend's pain, you also create a picture of his inner life. How upset is he? What is he thinking about? What will he do now? To answer these questions, you think like a detective, gathering evidence about his behavior and situation to deduce how he feels. This cognitive piece of empathy is referred to as "mentalizing," or explicitly considering someone else's perspective. Mentalizing, an everyday form of mind reading, is more sophisticated than experience sharing. It requires cognitive firepower that most animals don't have, and thus arrived later in evolution. And though children pick up experience sharing early, it takes them longer to sharpen their mentalizing skills.

CARING

If while your friend weeps, all you do is sit back, feel bad, and think about him, you're a less-than-stellar pal. Instead, you might also wish for him to feel better and hatch a plan for how you can get him there. This is what researchers call "empathic concern," or a motivation to improve someone else's well-being. This is the piece of empathy that most reliably sparks kind action. Concern has received less attention from Western researchers than mentalizing or experience sharing, though that is now changing. Concern also hews tightly to centuries-old formulations of "compassion" in the Buddhist tradition, for instance *karuna,* or the desire to free others from suffering.

SPLITS AND CONNECTIONS

Experience sharing, mentalizing, and concern split apart in all sorts of interesting ways. For instance, mentalizing is most useful when we *don't* share another's experiences. To know why a fan of a team you don't follow just climbed a signpost, you must understand differences between their emotional landscape and yours. When we fail to understand each other, it's often because we falsely assume our own knowledge or priorities will map onto someone else's.

Empathic processes activate different brain systems and are useful at different moments. Poker and boxing require keen mentalizing—What does your opponent know? What is her next move?—but are ill-served by concern. Parenting can be the opposite: You might never understand why your toddler is mid-meltdown, but you must still do what you can to help her. *People* also differ in their empathic landscapes. An emergency room physician likely feels great concern for her patients, but she cannot do her job if also taking on their pain. Individuals with autism spectrum disorder sometimes struggle at mentalizing, but they still share and care about others' feelings. Psychopaths have the opposite profile: They are perfectly able to tell what others feel but are unaffected by their pain.

Despite these distinctions, empathic pieces are also deeply intertwined. Sharing someone else's emotion draws our attention to what they feel, and thinking about them reliably increases our concern for their well-being. All three empathic processes promote kindness, albeit in distinct ways. The primatologist Frans de Waal describes this beautifully in his "Russian Doll Model" of empathy. As he sees it, the primitive process of experience sharing is at the core—turning someone else's pain into our own creates an impulse to stop it. Newer, more complex forms of empathy are layered on top of that, generating broader sorts of kindness. Through mentalizing, we develop a fine-grained picture of not just what someone else feels, but why they feel it, and—more important—what might make them feel better. This spurs a deeper concern, a response focused not only on our own discomfort but truly on someone else. The global kindness Peter Singer describes in *The Expanding Circle* is a further extension of concern—pointed not at any one individual, but at people as a whole.

This book focuses on rebuilding empathy when it's eroded. Pinpointing different pieces of empathy helps researchers diagnose what has gone wrong and find the most effective solutions. Callousness can come from thoughtlessness: We discount the suffering of a homeless person because we don't consider their experiences at all. In that case, interventions might focus on encouraging mentalizing, for instance, through perspective-taking exercises or virtual reality. In the face of conflict, we might think a great deal about our enemies but not care about their well-being (we may even hope for them to suffer). Contact, and especially friendships across group lines, can change that. Burnout—for instance, among medical professionals—often follows from *too much* experience sharing. Contemplative techniques can help people shift themselves toward concern instead. In all these cases, understanding what to do with empathy requires first understanding exactly what it is.

Evaluating the Evidence

AT VARIOUS TIMES, scientific texts confirmed that the sun revolved around the earth, atoms were the smallest particles in the universe, and the human soul could be located in the pineal gland. The scientific method allowed all of these "facts" to be overwritten as the truth came to light. It's this dynamism, and the humility that must accompany it, that gives science its power. Science is not a set of facts, but a process of predicting, testing, and rethinking. It is alive.

In this book, I review scientific evidence about the forces that strengthen and weaken human empathy and kindness. Most of this evidence comes from the field of psychology. Over the past several years, some high-profile psychological findings have proven less robust than they had seemed. Similar doubts have arisen about findings in political science, economics, biology, and medical research. We psychologists have used this as an opportunity to strengthen our methods, be more transparent about our research process, and clarify exactly what we do and don't know.

In that spirit, I decided that readers should have tools for further evaluating the evidence presented in this book. Some of the claims you've read are supported by decades of highly consistent research. Others are brand-new insights published for the first time in recent weeks or months. I would not have written about new findings if I did not believe in them, but it's impossible to know for certain what their status will be years from now.

This appendix is meant to convey how robust the evidence is underlying the claims in this book. To generate it, my colleague Kari Leibowitz (who also cowrote this section of the book with me) first gathered independent evidence on key claims in each chapter, and rated the strength of each claim from 5 (strongest) to 1 (weakest). She and I then convened to discuss, see if either of us had missed any evidence, and agree on all scores. If Kari and I deemed a claim to be only weakly supported, I either removed it from the text or clarified its tentative nature. In other words, this appendix doesn't merely evaluate the content of the book; it helped shape it.

THE RATING SYSTEM

Each claim rating consists of either two or three parts:

1. A summary of the claim, drawn from the book

2. A 1 to 5 rating (from weaker to stronger)

3. A brief description of the reasoning behind the rating (only if a claim received a rating of 1 to 3)

Most strong evidence is alike, but imperfect evidence tends to be imperfect in its own way. This is especially true of claims we rated as a 3, 2, or 1, which is why those claims further include a brief explanation of why they received the score they did. That said, here are guidelines to help you understand what each rating *generally* means.

5. Indicates that the evidence in support of this claim is extremely strong. This research has been replicated many times, meaning that independent researchers have almost all found evidence supporting the claim. In many cases, it means the research has been conducted with large numbers of diverse participants. Often, *meta-analyses* support this claim. In meta-analyses,

many studies are combined to see if the effect of a particular phenomenon is significant when assessed across all of them.

4. Indicates that the evidence in support of this claim is very strong. This research has been replicated multiple times, with evidence generally supporting the claim. The evidence here is just shy of being as strong as possible: This may be because this claim appears to hold in some but not all contexts, or because there haven't been large-scale meta-analyses on this topic.

3. Indicates fairly strong evidence. Here, the research has indeed been replicated, but perhaps the effect is observed only in some settings (for example, in schools but not workplaces), or the research hasn't been replicated with large numbers of diverse participants. A claim may also receive a rating of 3 if it is fairly robust but still new enough that there haven't been a large number of replications.

2. Indicates that there is not a large body of research supporting the claim. This could be because replications of this research have found conflicting results, or because the research is so new or labor intensive that it is difficult for independent researchers to conduct replications. Claims that receive a 2 could also represent findings that hold (or have been tested) only in highly specific settings.

1. Indicates that there is only a small amount of evidence supporting the claim. This could be because the research is new, or is so difficult to conduct that scientists have not been able to independently replicate it. A rating of 1 does not mean that the claim isn't true; rather, it means that there is not yet a great deal of evidence supporting it.

A high rating does not mean that a given effect is observed in all situations and with all people. By way of analogy, consider medicine. Even our most advanced, well-supported treatments don't work for all illnesses, or even for every person with the same illness. A medication's effectiveness relies on a complex interaction between its pharmaceutical properties, a patient's biology, and the details of their illness. Likewise, psychological effects reflect complex interactions between a person's psychology and their social context. The best psychological theories predict when, where, and why an effect will be observed.

The ratings in this appendix are meant to be bite-sized summaries of the strength of evidence for claims in this book. These are necessarily simplified, in most cases distilling decades of research into a single number. If you want more information than is provided here, you can find a spreadsheet containing the research that went into vetting each claim at www.warforkindness.com/data. I hope that this appendix and these resources will give you a glimpse into the work this book relies on. More broadly, I hope it will drive home how dynamic the science of human empathy and kindness is.

INTRODUCTION

Overall: The book begins by introducing some of the most foundational concepts of empathy, including its evolutionary roots, bases in the brain, and its relationship to kind actions. As such, the claims in this chapter rest on strong theoretical and empirical evidence, and are supported by a wealth of interdisciplinary research from fields including psychology, neuroscience, evolutionary biology, and economics.

Claim 0.1: Empathy is related to kindness and prosociality.
   Rating: 5

Claim 0.2: Evolution favors empathy, through selective advantages for prosocial organisms.
   Rating: 5

Claim 0.3: Empathic individuals excel professionally.
   Rating: 4

Claim 0.4: Empathic individuals experience greater subjective well-being.
   Rating: 4

Claim 0.5: It is easier to empathize with one person than many people.
   Rating: 4

Claim 0.6: "Mirroring" in the brain is associated with empathy.
Rating: 5

CHAPTER 1: THE SURPRISING MOBILITY OF HUMAN NATURE

Overall: Several claims in this chapter, such as the genetic and environmental determinants of empathy, are backed up by a great deal of interdisciplinary research. However, other claims concern research that is either so new it has yet to be replicated or claims for which a consensus is still emerging.

Claim 1.1: IQ/intelligence can change with experience.
Rating: 5

Claim 1.2: Empathy is, in part, genetically determined.
Rating: 5

Claim 1.3: Children's environments impact their levels of empathy.
Rating: 4

Claim 1.4: People who carry out necessary evils (such as giving bad news) experience reduced empathy.
Rating: 4

Claim 1.5: People who undergo intense suffering often become more prosocial as a result.
Rating: 3
There is ample evidence to support the claim that intense suffering can lead to increased empathy and prosociality. But in other cases, the opposite is true—violence begets violence, and suffering can make people crueler or abusive. Several review papers outline well-articulated theories of when and why suffering should lead to positive outcomes, rather than downward cycles, but further research is needed to test these theories.

Claim 1.6: Mindsets about the malleability of empathy influence people's empathy.
Rating: 1

Our work cited in this chapter is the first examining mindsets about empathy, and it has yet to be tested in many independent replication studies. Two projects have examined the construct of empathic mindsets in tangentially related ways. One found that people low in empathy tend toward aggression, but not if they hold a malleable mindset. Another found that inducing a malleable mindset did not increase the likelihood that participants would forgive prisoners for their crimes. Overall, more research on this topic is needed to confirm the effects of mindsets about empathy we documented.

CHAPTER 2: CHOOSING EMPATHY

Overall: This chapter considers people's control over their emotions and empathy, and highlights motives that drive people toward or away from empathizing. Several of these claims are based on years of well-established evidence, but others are still subject to debate among psychologists.

Claim 2.1: We have the ability to control and regulate our emotions.
  Rating: 5

Claim 2.2: In situations when it's important to build relationships, people ramp up their empathy.
  Rating: 3
Much research supports the idea that empathy is important for relationship building. Further, research on loneliness and impression management does suggest that people in these situations ramp up their empathy. But relatively little research explicitly creates situations in which relationship building is critical in order to causally test the effects of affiliative desire on empathy.

Claim 2.3: People empathize to help bolster their moral self-image.
  Rating: 4

Claim 2.4: When people think empathizing will be painful, they avoid it.
  Rating: 4

Claim 2.5: Stress reduces empathy.

Rating: 3

Particularly in caring professions where burnout is common, evidence supports a link between increased stress and burnout and reduced empathy. However, most studies on this topic are correlational, not causal. Furthermore, although some studies demonstrate that stress can reduce perspective-taking abilities, at least one study finds that stress *can increase* short-term prosocial behavior. More experimental studies are needed to fully understand the relationship between stress and empathy.

Claim 2.6: When people believe empathy is a valued norm, they empathize more.

Rating: 4

Claim 2.7: Purposely cultivating empathy alters the brain.

Rating: 3

Several well-conducted studies indicate that empathy and compassion training lead to corresponding changes in the brain. However, almost all of this work focuses on brain changes resulting from contemplative practice, such as loving-kindness meditation. These studies should be augmented by additional research examining the neural effects of other empathy-building practices.

CHAPTER 3: HATRED VERSUS CONTACT

Overall: Centered around contact theory and its related claims, this chapter's foundation is one of the most well-studied areas in the social sciences. The majority of claims in this chapter have been tested in diverse contexts, across hundreds of studies with thousands of participants.

Claim 3.1: People naturally empathize more with members of their ingroups, as compared to outsiders.

Rating: 5

Claim 3.2: We fail to empathize—and often experience antipathy—in competitive contexts.

Rating: 5

Claim 3.3: Contact generally increases empathy for outsiders.

Rating: 5

Claim 3.4: Contact can bolster empathy for outsiders amid conflict or competition.

Rating: 5

Claim 3.5: Specific conditions (for example, those laid out by Gordon Allport) are necessary in order for contact to foster empathy toward outsiders

Rating: 3

There are many studies documenting differences in the effects of contact across situations, but little agreement about which parameters are required for contact to "work," as reflected in a recent meta-analysis.

CHAPTER 4: THE STORIES WE TELL

Overall: This chapter discusses the role of narrative arts in building empathy. Compared to contact theory, there are relatively few well-controlled studies assessing the impact of the arts on our empathy. Claims in this chapter receive a low score based not on negative findings, but rather on a lack of systematic replication. That said, an increasing number of studies—including a recent meta-analysis—provide emerging evidence for the effects of storytelling on empathy. More research in this domain will help confirm or contextualize these findings.

Claim 4.1: Theater grows empathy.

Rating: 3

Despite a few well-conducted studies with promising findings, other work in this domain often a) relies on self-report; b) leads to no significant objective improvements; or c) has no control group. More well-designed, empirical work is needed to examine the extent to which theater practices grow empathy.

Claim 4.2: Literature grows empathy.
   Rating: 4

Claim 4.3: Reading literature can reduce criminal offenses.
   Rating: 1
Much research exists showing the benefits of education for incarcerated individuals, and there are many anecdotal reports of the benefits of bibliotherapy (that is, reading literature) for prison populations. However, outside of the evaluation of Changing Lives Through Literature itself, there are almost no experimental tests of the benefit of reading literature on the outcome of criminal recidivism.

Claim 4.4: Narrative art can reduce intergroup conflict.
   Rating: 4

CHAPTER 5: CARING TOO MUCH
Overall: This chapter highlights the benefits and limitations of empathy in the context of caregiving, with a focus on medical settings. The majority of claims rest on well-documented, rigorous research, including large-scale randomized control trials and meta-analyses, though one claim is also the subject of some debate.

Claim 5.1: Compassion fatigue is prevalent among caring professions and detrimental to them.
   Rating: 5

Claim 5.2: Provider empathy has salutary consequences for patient outcomes.
   Rating: 5

Claim 5.3: For healthcare professionals, empathy has pitfalls.
   Rating: 3

The findings around this claim are mixed: Evidence suggests empathy can produce negative consequences for medical professionals—including burnout, distress, and reduced provider efficacy—but other research suggests empathy can protect against burnout and increase provider efficacy. As described in the chapter, this likely hinges on the *type* of empathy (distress versus concern) caregivers experience.

Claim 5.4: Social support buffers against burnout.
Rating: 5

Claim 5.5: Mindfulness reduces burnout for caregivers.
Rating: 5

Claim 5.6: Mindfulness increases caregiver empathy.
Rating: 4

CHAPTER 6: KIND SYSTEMS

Overall: Some of this chapter's claims, such as those concerning the power of norms, are extremely well supported by decades of interdisciplinary research. Other claims, however, such as those involving the potential for empathy interventions to transform policing and classroom discipline, rely on new research that has yet to be replicated on a large scale.

Claim 6.1: Social norms influence our thoughts and actions.
Rating: 5

Claim 6.2: People conform to perceived norms and often overestimate the prevalence of extreme positions.
Rating: 5

Claim 6.3: Empathy begets empathy: Positive and empathic norms spread.
Rating: 4

Claim 6.4: Interpersonal training programs for police improve policing outcomes.
Rating: 3

Several studies suggest that training focused on police officer empathy, conflict management, or procedural justice can improve policing outcomes, for instance, by helping officers de-escalate dangerous situations. However, relatively few studies have assessed such training programs experimentally, using control groups and looking at important outcomes over time.

Claim 6.5: Empathy bias, or preferential empathy for one's in-group, often outweighs an individual's overall empathy, particularly during intergroup conflict.

Rating: 2

The study by Bruneau et al. cited with this claim is quite recent and has not yet been replicated by multiple, independent groups. Although it is consistent with many studies on empathy bias and the power of in-group/out-group empathy, there have been almost no studies directly comparing empathy bias to overall empathy. Thus, the specific argument that empathy bias matters more than overall empathy has not been empirically tested in many studies.

Claim 6.6: Social and Emotional Learning programs lead to many benefits (particularly for young children).

Rating: 5

Claim 6.7: Empathic discipline helps classrooms.

Rating: 1

Jason Okonofua's study on this topic is promising and well conducted, but this research is so new it has yet to be replicated, and no other studies have examined the impact of empathy-focused discipline in an educational setting.

CHAPTER 7: THE DIGITAL DOUBLE EDGE

Overall: This chapter's central claims concern the impact of technology on human empathy. Like the technology discussed, much of this research is new. However, many of the claims in this chapter—about both the positive and negative effects of technology on empathy—are well supported. The most controversial claim of this chapter, accordingly, is whether increased technology and Internet use increases or decreases empathy. As the other claims indicate: It depends on how we use it!

Claim 7.1: Increased technology/Internet use is associated with decreases in empathy.
Rating: 2
When technology and Internet use supplant face-to-face interaction, they leave people less sociable and less likely to interact with those around them. However, in other cases, Internet and social media use can make people more empathetic and open-minded. In other words, online experiences can either increase or decrease empathy, depending on whether they replace or supplement other types of social interaction.

Claim 7.2: Internet anonymity encourages cyberbullying.
Rating: 4

Claim 7.3: Internet echo chambers encourage and reward extreme and emotional views.
Rating: 4

Claim 7.4: Virtual reality experiences can decrease stereotyping and discrimination.
Rating: 4

Claim 7.5: Virtual reality can build empathy.
Rating: 4

Claim 7.6: Online communities can provide meaningful and helpful support to their members.
Rating: 4

Claim 7.7: Giving to others helps the helper, making them happier or more fulfilled.
Rating: 5

Overall: Claims in the epilogue focus on leveraging empathy and related emotions to create a better future. While some of this research is well supported, much of it is new. More direct replications will help assess the reliability of this early work.

Claim 8.1: Gratitude helps people think about the long term.

Rating: 3

Several promising studies suggest that gratitude helps facilitate long-term thinking, making people more likely to delay gratification or sacrifice for future generations. However, this body of evidence is still relatively small, and further research is needed to replicate these findings.

Claim 8.2: Awe inspires connectedness and generosity.

Rating: 4

Claim 8.3: When people think about their legacy, they become more committed to sacrificing for future generations.

Rating: 3

Several compelling studies suggest that helping individuals think about their legacies encourages people to sacrifice for future generations, for example, by engaging in sustainable activities to address climate change. However, these lines of research are relatively new and need to be replicated.

INTRODUCTION

4   **we get it right:** For more on accuracy in empathic judgment, see
Jamil Zaki and Kevin N. Ochsner, "Reintegrating Accuracy into the
Study of Social Cognition," *Psychological Inquiry* 22, no. 3 (2011):
159–82.

4   **Glimpsing a face:** R. Thora Bjornsdottir and Nicholas O. Rule, "The
Visibility of Social Class from Facial Cues," *Journal of Personality
and Social Psychology* 113, no. 4 (2017): 530; John Paul Wilson and
Nicholas O. Rule, "Advances in Understanding the Detectability of
Trustworthiness from the Face," *Current Directions in Psychological
Science* 26, no. 4 (2017): 396–400; and Michael S. North et al., "In-
ferring the Preferences of Others from Spontaneous, Low-Emotional
Facial Expressions," *Journal of Experimental Social Psychology* 46,
no. 6 (2010): 1109–13.

4   **to inspire kindness:** Like "empathy," "kindness" can be divided into
a number of categories. "Prosociality" refers to behavior that helps
others. Within that category, "cooperation" refers to mutual aid, such
as working together toward a goal that benefits all parties. "Altruism," by contrast, refers to acts that help recipients at no benefit—or
at a cost—to helpers. Cooperation and altruism are often surprisingly
difficult to tell apart, because people can benefit from helping others

in all sorts of ways, such as winning admiration or even simply feeling good. A review of these terms is provided by Stuart A. West et al., "Social Semantics: Altruism, Cooperation, Mutualism, Strong Reciprocity and Group Selection," *Journal of Evolutionary Biology* 20, no. 2 (2007): 415–32.

4   **As Darwin wrote:** Charles Darwin, *The Descent of Man, and Selection in Relation to Sex* (London: John Murray, 1871).

5   **Empathy is nature's answer:** For more on the evolutionary role of empathy in supporting kindness across the animal kingdom, see Frans de Waal, "Putting the Altruism Back into Altruism: The Evolution of Empathy," *Annual Review of Psychology* 59 (2008): 279–300; Stephanie Preston, "The Origins of Altruism in Offspring Care," *Psychological Bulletin* 139, no. 6 (2013): 1305–41; and Jean Decety et al., "Empathy as a Driver of Prosocial Behavior: Highly Conserved Neurobehavioral Mechanisms Across Species," *Philosophical Transactions of the Royal Society B: Biological Sciences* 371, no. 1686 (2016): 52–68.

5   **Mice, elephants, monkeys, and ravens:** See de Waal, "Putting the Altruism Back into Altruism"; Decety, "Empathy as a Driver of Prosocial Behavior"; and Jeffrey S. Mogil, "The Surprising Empathic Abilities of Rodents," *Trends in Cognitive Sciences* 16, no. 3 (2012): 143–44.

5   **five other large-brained human species:** Yuval Noah Hariri, *Sapiens: A Brief History of Humankind* (New York: HarperCollins, 2017).

5   ***sapiens* changed to make connecting easier:** Brian Hare, "Survival of the Friendliest: *Homo sapiens* Evolved Via Selection for Prosociality," *Annual Review of Psychology* 68 (2017): 155–86; Robert L. Cieri et al., "Craniofacial Feminization, Social Tolerance, and the Origins of Behavioral Modernity," *Current Anthropology* 55, no. 4 (2014): 419–43; and Michael Tomasello, *A Natural History of Human Morality* (Cambridge, Mass.: Harvard University Press, 2016). Cieri et al. describe "behavioral modernity," including use of tools, projectile weapons, artistic expression, and long-distance communication. Behavioral modernity arose only about fifty thousand years ago, well

after our brains had developed to near-contemporary size and structure. Why? Potentially because increased population density required us to develop cooperative strategies for hunting and protection, and thus reduced our aggression.

6 **This became our secret weapon:** One explanation for humanity's meteoric rise is what's known as "cultural ratcheting," people's ability to transmit innovations such that each generation builds on the last, which in turn depends on our ability to care for and understand one another. See, for instance, Claudio Tennie et al., "Ratcheting Up the Ratchet: On the Evolution of Cumulative Culture," *Philosophical Transactions of the Royal Society B: Biological Sciences* 364, no. 1528 (2009): 2405–15.

6 **Americans alone donated $410 billion:** *Donations: Giving USA 2018: The Annual Report on Philanthropy for the Year 2017,* a publication of the Giving USA Foundation, 2017, researched and written by the Indiana University Lilly Family School of Philanthropy; available online at www.givingusa.org. *Volunteering: Volunteering and Civic Life in America,* Corporation for National and Community Service, retrieved from www.nationalservice.gov on July 16, 2018.

6 **Highly empathic individuals donate more:** C. Daniel Batson, *Altruism in Humans* (Oxford: Oxford University Press, 2011); and Nancy Eisenberg and Richard A. Fabes, "Empathy: Conceptualization, Measurement, and Relation to Prosocial Behavior," *Motivation and Emotion* 14, no. 2 (1990): 131–49.

Of course, empathy is not the *only* force that drives kindness. It can also be spurred by obligation, laws, or what the economist René Bekkers calls the "principle of care": a universal ideal that people should help each other. Bekkers and his colleagues have demonstrated that empathic people are more likely to believe in a principle of care, and that principle, in turn, drives their helping. For more on this, see René Bekkers and Mark Ottoni-Wilhelm, "Principle of Care and Giving to Help People in Need," *European Journal of Personality* 30, no. 3 (2016): 240–57. There is very little research on extraordinary acts of kindness, but at least some now indicates that it, too, might be supported by empathy. For instance, altruistic kidney donors—

who donate to strangers—exhibit a greater degree of brain "mirroring" in response to others' distress than nondonors. See Kristin M. Brethel-Haurwitz et al., "Extraordinary Altruists Exhibit Enhanced Self-Other Overlap in Neural Responses to Distress," *Psychological Science* 29, no. 10 (2018): 1631–41.

6   ***The Expanding Circle:*** Peter Singer, *The Expanding Circle: Ethics, Evolution, and Moral Progress* (Princeton, N.J.: Princeton University Press, 2011).

7   **hormones that encourage parents:** For instance, oxytocin is involved in maternal behaviors, and often thought of as a general "love hormone." But it can also inspire parochialism: kindness toward one's own group, but aggression toward or exclusion of outsiders. See, for instance, Carsten De Dreu et al., "Oxytocin Promotes Human Ethnocentrism," *Proceedings of the National Academy of Sciences* 108, no. 4 (2011): 1262–66.

7   **more people lived in cities:** United Nations, World Urbanization Prospects, 2018 Revision.

7   **we are increasingly isolated:** K. D. M. Snell, "The Rise of Living Alone and Loneliness in History," *Social History* 42, no. 1 (2017): 2–28.

8   **The average person in 2009:** Sara Konrath et al., "Changes in Dispositional Empathy in American College Students over Time: A Meta-Analysis," *Personality and Social Psychology Review* 15, no. 2 (2011): 180–98. The 75 percent figure pertains to empathic concern (see appendix A for more on components of empathy); for perspective taking, a 2009 participant would be less empathic than about two-thirds of 1979 participants.

9   **"not the sheer size of the catastrophe":** Anne Barnard and Karam Shoumali, "Image of a Small, Still Boy Brings a Global Crisis into Focus," *New York Times,* September 3, 2015. Alan Kurdi's death might have even affected global politics. Canadian prime minister Stephen Harper called the images "heart-wrenching," expressing regret that the Kurdis had been unable to obtain visas to enter Canada. "You don't get to suddenly discover compassion in the middle of an elec-

tion campaign," shot back minority leader Justin Trudeau. Trudeau's party soon swept out Harper's. See Ian Austen, "Aylan Kurdi's Death Resonates in Canadian Election Campaign," *New York Times,* September 4, 2014.

9 **our compassion collapses:** Paul Slovich and other psychologists term this the "identifiable victim effect" (IVE), building on the ideas of Thomas Schelling. For more, see Seyoung Lee and Thomas H. Feeley, "The Identifiable Victim Effect: A Meta-Analytic Review," *Social Influence* 11, no. 3 (2016): 199–215. For the IVE as it pertains to Kurdi's case, see Paul Slovicet et al., "Iconic Photographs and the Ebb and Flow of Empathic Response to Humanitarian Disasters," *Proceedings of the National Academy of Sciences* 114, no. 4 (2017): 640–44.

9 **"There's a lot of talk":** Barack Obama, 2006 Northwestern Commencement [transcript], retrieved from https://www.northwestern .edu/newscenter/stories/2006/06/barack.html.

10 **"The most important question":** Jeremy Rifkin, *The Empathic Civilization: The Race to Global Consciousness in a World in Crisis* (New York: Penguin, 2009).

11 **Francis Galton, a British scientist:** Francis Galton, "Hereditary Talent and Character," *Macmillan's Magazine* 12, nos. 157–66 (1865): 318–27; Raymond E. Fancher, "Biographical Origins of Francis Galton's Psychology," *Isis* 74, no. 2 (1983): 227–33; and Arthur R. Jensen, "Galton's Legacy to Research on Intelligence," *Journal of Biosocial Science* 34, no. 2 (2002): 145–72. Note that Galton's "Anthropometric Laboratory" tested not only capacities related to intelligence but also more purely physical attributes, such as strength of people's punches.

12 **dozens of assessments:** A smattering of these throughout the ages: Edwin G. Boring and Edward Titchener, "A Model for the Demonstration of Facial Expression," *American Journal of Psychology* 34, no. 4 (1923): 471–85; Dallas E. Buzby, "The Interpretation of Facial Expression," *American Journal of Psychology* 35, no. 4 (1924): 602–4; Rosalind Dymond, "A Scale for the Measurement of Empathic Ability," *Journal of Consulting Psychology* 13, no. 2 (1949): 127–33; Robert Rosenthal et al., *Sensitivity to Nonverbal Communication: The PONS*

Test (Baltimore, Md.: Johns Hopkins University Press, 1979); Simon Baron-Cohen et al., "The 'Reading the Mind in the Eyes' Test Revised Version: A Study with Normal Adults, and Adults with Asperger Syndrome or High-Functioning Autism," *Journal of Child Psychology and Psychiatry* 42, no. 2 (2001): 241–51; and Ian Dziobek et al., "Introducing MASC: A Movie for the Assessment of Social Cognition," *Journal of Autism and Developmental Disorders* 36, no. 5 (2006): 623–36.

12   **How much did someone's heart rate jump**: In *Blade Runner,* one key method for telling the difference between people and replicants— human-looking robots who don't know they're synthetic—boils down to an empathy test of this sort. In the Voight-Kampff, subjects are shown evocative images, including those of others in pain. If their palms sweat, they're human; if not, they're a replicant.

12   **findings were less straightforward:** For instance, empathic concern, cognitive empathy, and emotional empathy are only weakly to moderately correlated across people; see Mark Davis, "Measuring Individual Differences in Empathy: Evidence for a Multidimensional Approach," *Journal of Personality and Social Psychology* 44, no. 1 (1983): 113–26. For evidence that some, but not all, empathy tests predict prosocial behavior, see Nancy Eisenberg and Paul A. Miller, "The Relation of Empathy to Prosocial and Related Behaviors," *Psychological Bulletin* 101, no. 1 (1987): 91–119.

12   **emotional intelligence (EI):** EI encompasses empathy but also other abilities, most important, how clearly someone understands and regulates their *own* emotion. For more, see Peter Salovey and John D. Mayer, "Emotional Intelligence," *Imagination, Cognition and Personality* 9, no. 3 (1990): 185–211; and John D. Mayer et al., "The Ability Model of Emotional Intelligence: Principles and Updates," *Emotion Review* 8, no. 4 (2016): 290–300. For a critique of EI, see Gerald Matthews et al., "Seven Myths About Emotional Intelligence," *Psychological Inquiry* 15, no. 3 (2004): 179–96.

12   **Data's hijinks:** One exception is the string of *Star Trek: The Next Generation* episodes in which Data's creator, Noonien Soong, gives him an "emotion chip." But we don't have to dwell on those.

13   **emotions are "contagious":** For more on the idea of emotion contagion, both historical and scientific, see Adam Smith, *The Theory of Moral Sentiments* (Cambridge: Cambridge University Press, 2002; first published in 1790); Gustav Jahoda, "Theodor Lipps and the Shift from 'Sympathy' to 'Empathy,'" *Journal of the History of the Behavioral Sciences* 41, no. 2 (2005): 151–63; Elaine Hatfield et al., *Emotional Contagion* (Cambridge: Cambridge University Press, 1994); and Edith Stein, *On the Problem of Empathy* (Washington, D.C.: ICS, 1989; first published in English in 1964).

    Edith Stein's personal story is as interesting as her philosophy. Though she greatly advanced the study of empathy, as a woman she could not lecture, so was relegated to being Edmund Husserl's personal assistant. She quit after converting from Judaism to atheism to Catholicism, and became a nun in a Dutch convent. She nonetheless was rounded up as a Jew by Nazi forces and died alongside her sister in Auschwitz. Four decades later, she was canonized by Pope John Paul II as a saint. For more, see Alisdair MacIntyre, *Edith Stein: A Philosophical Prologue, 1913–1922* (Lanham, Md.: Rowman and Littlefield, 2007).

13   **"mirror neurons":** The original papers in this series include Giuseppe di Pellegrino et al., "Understanding Motor Events: A Neurophysiological Study," *Experimental Brain Research* 91, no. 1 (1992): 176–80; and Vittorio Gallese et al., "Action Recognition in the Premotor Cortex," *Brain* 119, no. 2 (1996): 593–609.

14   **documented human mirroring:** For review, see, for example, Christian Keysers and Valeria Gazzola, "Expanding the Mirror: Vicarious Activity for Actions, Emotions, and Sensations," *Current Opinion in Neurobiology* 19, no. 6 (2009): 666–71; Claus Lamm et al., "Meta-Analytic Evidence for Common and Distinct Neural Networks Associated with Directly Experienced Pain and Empathy for Pain," *Neuroimage* 54, no. 3 (2011): 2492–502; and Sylvia S. Morelli et al., "Common and Distinct Neural Correlates of Personal and Vicarious Reward: A Quantitative Meta-Analysis," *Neuroimage* 112 (2014): 244–53.

14   **seemed to inspire kindness:** Grit Hein et al., "Neural Responses to Ingroup and Outgroup Members' Suffering Predict Individual Differences in Costly Helping," *Neuron* 68, no. 1 (2010): 149–60;

and Jamil Zaki et al., "Activity in Ventromedial Prefrontal Cortex Covaries with Revealed Social Preferences: Evidence for Person-Invariant Value," *Social Cognitive Affective Neuroscience* 9, no. 4 (2014): 464–69.

14   **mirroring failed to predict:** Lamm et al., "Meta-Analytic Evidence for Common and Distinct Neural Networks," for instance, report that about 60 percent of studies that measure brain mirroring of pain also find correlations between mirroring and subjective empathy (see page 2500).

14   **not clear exactly how mirroring works:** Evidence overwhelmingly suggests that people produce overlapping patterns of brain activity when observing emotional (and other) states in others and when experiencing them directly. Does that mean that personal and vicarious emotions are identical? No. If they were, we'd live in a confusing world, unable to tell the difference between ourselves and everyone else. Many brain regions are activated by all sorts of different experience. For instance, parts of the brain's frontal lobe respond to both accessing memories and generating speech. That doesn't imply that talking and remembering are the same thing.

    Psychological states usually can't be pulled out from the activity of any single part of the brain; instead they reflect *patterns* of activity across the brain. At the level of these patterns, personal and vicarious experiences can be distinguished. The takeaway as I see it is that *mirroring does not mean that personal and empathic experience are exactly the same, but suggests that they share some key features.* For more on this perspective, see Jamil Zaki et al., "The Anatomy of Suffering: Understanding the Relationship Between Nociceptive and Empathic Pain," *Trends in Cognitive Sciences* 20, no. 4 (2016): 249–59.

14   **"Gandhi neurons":** Vilayanur S. Ramachandran, "The Neurons That Shaped Civilization," talk delivered at TEDIndia, 2009.

14   **Brain images . . . evoke the truth:** For instance, people are more likely to believe even dubious claims about psychological processes if they're accompanied by images of brain scans. See, for example, Dana S. Weisberg et al., "Deconstructing the Seductive Allure of

Neuroscience Explanations," *Judgment and Decision Making* 10, no. 5 (2015): 429.

14   **minds as "hardwired":** This term originates from Carl Sagan's *The Dragons of Eden.* In it, Sagan characterizes one view of the brain as "completely hard-wired: specific cognitive functions are localized in particular places in the brain." He goes on to argue that hard wiring is likely *not* the right metaphor for neuroscience, but as fMRI leapt into our collective consciousness, hard-wiring tagged along for better and for worse. See Carl Sagan, *Dragons of Eden: Speculations on the Evolution of Human Intelligence* (New York: Ballantine, 2012; first published in 1977).

16   **As the novelist George Saunders writes:** Joel Lovell, "George Saunders's Advice to Graduates," *New York Times,* July 31, 2013.

CHAPTER 1: THE SURPRISING MOBILITY OF HUMAN NATURE

17   **Alfred Wegener changed that:** Details on Wegener's work and life are drawn from Martin Schwarzbach, *Alfred Wegener: The Father of Continental Drift* (Madison, Wis.: Science Tech, 1986); and Anthony Hallam, *Great Geological Controversies* (Oxford: Oxford University Press, 1989).

17   **he wrote to a lady friend:** His pen pal, Else Köppen, later married Wegener. It's unclear whether his geographical poetics played any role in wooing her.

19   **defending prevailing social hierarchies:** In addition to phrenology, other kinds of biological determinism have been trotted out to defend racial, gender, and class hierarchies over the centuries. For a review, see Stephen Jay Gould, *The Mismeasure of Man* (New York: W. W. Norton, 1996).

19   **The father of modern neuroscience:** Santiago Ramón y Cajal, *Estudios Sobre la Degeneración y Regeneración del Sistema Nervioso* (Madrid: Moya, 1913).

20   **thousands of new brain cells:** Arturo Alvarez-Buylla et al., "Birth of Projection Neurons in the Higher Vocal Center of the Canary

Forebrain Before, During, and After Song Learning," *Proceedings of the National Academy of Sciences* 85, no. 22 (1988): 8722–26. For a review of neuroplasticity in non-human animals, see Charles Gross, "Neurogenesis in the Adult Brain: Death of a Dogma," *Nature Reviews Neuroscience* 1, no. 1 (2000): 67–73.

20   **people grow new neurons:** See, for instance, Kirsty L. Spalding et al., "Dynamics of Hippocampal Neurogenesis in Adult Humans," *Cell* 153, no. 6 (2013): 1219–27. There is some controversy about the *amount* of new cells produced by the adult brain. In April 2018, one group reported negligible growth of new cells in the hippocampus after childhood, but just a month later another reported evidence that the same brain region continues producing new cells even in older adults. See Shawn F. Sorrells et al., "Human Hippocampal Neurogenesis Drops Sharply in Children to Undetectable Levels in Adults," *Nature* 555 (2018): 377–81; and Maura Boldrini et al., "Human Hippocampal Neurogenesis Persists Throughout Aging," *Cell Stem Cell* 22, no. 4 (2018): 589–99. But the preponderance of evidence suggests that adults do form at least *some* new neurons.

20   **habits mold our brains:** *Stringed instruments:* Thomas Elbert et al., "Increased Cortical Representation of the Fingers of the Left Hand in String Players," *Science* 270, no. 5234 (1995): 305–7; *juggling:* Bogdan Draganski et al., "Neuroplasticity: Changes in Grey Matter Induced by Training," *Nature* 427, no. 6972 (2004): 311–12; **stress:** Robert M. Sapolsky, "Glucocorticoids and Hippocampal Atrophy in Neuropsychiatric Disorders," *Archives of General Psychiatry* 57, no. 10 (2000): 925–35. These studies typically measure the volume of gray matter, and it's important to note that volumetric increase does not necessarily imply the creation of new cells. It can also arise, for instance, through increasing numbers of connections *between* cells. For more on "behavior-dependent plasticity," see Alvaro Pascual-Leone et al., "The Plastic Human Brain Cortex," *Annual Review of Neuroscience* 28 (2005): 377–401.

21   **IQ had shot up:** James R. Flynn, "Massive IQ Gains in 14 Nations: What IQ Tests Really Measure," *Psychological Bulletin* 101, no. 2 (1987): 171–91. For evidence of IQ change within families, see Bernt

Bratsberg and Ole Rogeberg, "Flynn Effect and Its Reversal Are Both Environmentally Caused," *Proceedings of the National Academy of Sciences* 155, no. 26 (2018): 6674–78. Interestingly (and troublingly), these authors found that intelligence—at least in Norway—has *dropped* in recent years. Both earlier increases and these drops appear to be environmentally mediated.

21   **reflect new choices and habits:** For a review highlighting environmental effects of intelligence, including through adoption, see Richard E. Nisbett et al., "Intelligence: New Findings and Theoretical Developments," *American Psychologist* 67, no. 2 (2012): 130–59. For education effects, see Stuart J. Ritchie and Elliot M. Tucker-Drob, "How Much Does Education Improve Intelligence? A Meta-Analysis," *Psychological Science* 29, no. 8 (2018): 1358–69. It is possible that education might not *make* people smarter, but instead, people who are already smart might gravitate toward education. Ritchie and Tucker-Drob cleverly got around this problem by examining effects of schooling that cannot be explained this way: for instance, the introduction of compulsory schooling in new communities.

21   **Our personalities also change:** Daniel A. Briley and Elliot M. Tucker-Drob, "Genetic and Environmental Continuity in Personality Development: A Meta-Analysis," *Psychological Bulletin* 140, no. 5 (2014): 1303–31; Jule Specht et al., "Stability and Change of Personality Across the Life Course: The Impact of Age and Major Life Events on Mean-Level and Rank-Order Stability of the Big Five," *Journal of Personality and Social Psychology* 101, no. 4 (2011): 862–82; and Brent W. Roberts et al., "A Systematic Review of Personality Trait Change Through Intervention," *Psychological Bulletin* 143, no. 2 (2017): 117–41.

23   **empathy is about 30 percent genetically determined:** Ariel Knafo and Florina Uzefosky, "Variation in Empathy: The Interplay of Genetic and Environmental Factors," in *The Infant Mind: Origins of the Social Brain*, ed. Maria Legerstee et al. (New York: Guilford, 2013); and Salomon Israel et al., "The Genetics of Morality and Prosociality," *Current Opinion in Psychology* 6 (2015): 55–59. Heritability is a complex concept. Importantly (if not intuitively), the fact that

empathy is 30 percent genetically determined in these studies does not mean that it is exactly 70 percent environmentally determined, because genes and experiences interact in complex ways to create differences between people.

23   **comparable to IQ's genetic component:** The heritability of intelligence actually shifts throughout people's lifetime. Young children exhibit about 20 percent heritability, adults about 60 percent, and older adults about 80 percent. See Robert Plomin and Ian Deary, "Genetics and Intelligence Differences: Five Special Findings," *Molecular Psychiatry* 20, no. 1 (2015): 98–108.

23   **If you knew a person's score:** Daniel Grühn et al., "Empathy Across the Adult Lifespan: Longitudinal and Experience-Sampling Findings," *Emotion* 8, no. 6 (2008): 753–65.

24   **empathy is shaped by experience:** For a recent review, see Tracy L. Spinrad and Diana E. Gal, "Fostering Prosocial Behavior and Empathy in Young Children," *Current Opinion in Psychology* 20 (2018): 40–44. For specific studies cited here, see Amanda J. Moreno et al., "Relational and Individual Resources as Predictors of Empathy in Early Childhood," *Social Development* 17, no. 3 (2008): 613–37; Darcia Narvaez et al., "The Evolved Development Niche: Longitudinal Effects of Caregiving Practices on Early Childhood Psychosocial Development," *Early Childhood Research Quarterly* 28, no. 4 (2013): 759–73; Brad M. Farrant et al., "Empathy, Perspective Taking and Prosocial Behaviour: The Importance of Parenting Practices," *Infant and Child Development* 21, no. 2 (2012): 175–88; and Zoe E. Taylor et al., "The Relations of Ego-Resiliency and Emotion Socialization to the Development of Empathy and Prosocial Behavior Across Early Childhood," *Emotion* 13, no. 5 (2013): 822–31. Some of these studies control for parents' empathy, mitigating concerns that empathic parents' genes determine kids' empathy, with environment merely along for the ride.

24   **These children are spared:** Kathryn L. Humphreys et al., "High-Quality Foster Care Mitigates Callous-Unemotional Traits Following Early Deprivation in Boys: A Randomized Controlled Trial," *Jour-

nal of the American Academy of Child and Adolescent Psychiatry 54, no. 12 (2015): 977–83.

24   **A bout of depression:** Grühn, "Empathy Across the Adult Lifespan."

24   **We can't always avoid inflicting pain:** In one survey, 74 percent of oncologists reported delivering bad news at least five times a month. See Walter F. Baile et al., "SPIKES—A Six-Step Protocol for Delivering Bad News: Application to the Patient with Cancer," *Oncologist* 5, no. 4 (2000): 302–11. Layoff figures are based on 418,000 total in the United States in 2017, calculated by the firm Challenger, Gray, and Christmas.

24   **"necessary evils":** Joshua D. Margolis and Andrew Molinsky, "Navigating the Bind of Necessary Evils: Psychological Engagement and the Production of Interpersonally Sensitive Behavior," *Academy of Management Journal* 51, no. 5 (2008): 847–72.

24   **oncologists report feeling intense heartbreak:** Walter F. Baile, "Giving Bad News," *Oncologist* 20, no. 8 (2015): 852–53; and Robert L. Hulsman et al., "How Stressful Is Doctor–Patient Communication? Physiological and Psychological Stress of Medical Students in Simulated History Taking and Bad-News Consultations," *International Journal of Psychophysiology* 77, no. 1 (2010): 26–34.

25   **managers who swing the ax:** Leon Grunberg et al., "Managers' Reactions to Implementing Layoffs: Relationship to Health Problems and Withdrawal Behaviors," *Human Resource Management* 45, no. 2 (2006): 159–78.

25   **individuals who performed necessary evils:** Margolis and Molinsky, "Navigating the Bind of Necessary Evils."

25   **"moral disengagement":** Here I discuss disengagement as a *result* of violence, but it is also a *pathway* to cruelty and callousness. For a magisterial review of this phenomenon, see Albert Bandura, *Moral Disengagement: How People Do Harm and Live with Themselves* (New York: Worth, 2016).

25   **psychologists asked subjects to shock:** David C. Glass, "Changes in Liking as a Means of Reducing Cognitive Discrepancies Between

Self-Esteem and Aggression," *Journal of Personality* 32, no. 4 (1964): 531–49.

25   **doubted that Native Americans could feel:** Emmanuel Castano and Roger Giner-Sorolla, "Not Quite Human: Infrahumanization in Response to Collective Responsibility for Intergroup Killing," *Journal of Personality and Social Psychology* 90, no. 5 (2006): 804–18.

25   **they turn off:** Ervin Staub, *The Roots of Evil: The Origins of Genocide and Other Group Violence* (Cambridge: Cambridge University Press, 1989), 82.

25   **death workers at prisons:** Michael J. Osofsky et al., "The Role of Moral Disengagement in the Execution Process," *Law and Human Behavior* 29, no. 4 (2005): 371–93.

26   **Six months after:** Edna B. Foa and Barbara O. Rothbaum, *Treating the Trauma of Rape: Cognitive-Behavioral Therapy for PTSD* (New York: Guilford, 2001); and George Bonanno, "Loss, Trauma, and Human Resilience: Have We Underestimated the Human Capacity to Thrive After Extremely Aversive Events?" *American Psychologist* 59, no. 1 (2004): 20–28.

26   **Trauma survivors who are supported:** See, for example, Mary P. Koss and Aurelio J. Figueredo, "Change in Cognitive Mediators of Rape's Impact on Psychosocial Health Across 2 Years of Recovery," *Journal of Consulting and Clinical Psychology* 72, no. 6 (2004): 1063–72.

26   **"altruism born of suffering":** For more, see Johanna R. Vollhardt, "Altruism Born of Suffering and Prosocial Behavior Following Adverse Life Events: A Review and Conceptualization," *Social Justice Research* 22, no. 1 (2009): 53–97; and David M. Greenberg et al., "Elevated Empathy in Adults Following Childhood Trauma," *PLoS One* 13, no. 10 (2018).

Of course, not *all* trauma survivors become more altruistic, as evidenced by well-known "cycles of abuse" across generations, and the prevalence of childhood abuse among hate group members discussed in chapter 3: Hatred Versus Contact. But Vollhardt makes the important point that though trauma is often stereotyped as dooming its victims to cruelty, it often does the opposite.

26   **war-torn communities:** Michal Bauer et al., "Can War Foster Cooperation?," *Journal of Economic Perspectives* 30, no. 3 (2016): 249–74.

26   **80 percent of rape survivors:** Patricia Frazier et al., "Positive and Negative Life Changes Following Sexual Assault," *Journal of Consulting and Clinical Psychology* 69, no. 6 (2001): 1048–55.

27   **Participants stepped in to help:** Daniel Lim and David DeSteno, "Suffering and Compassion: The Links Among Adverse Life Experiences, Empathy, Compassion, and Prosocial Behavior," *Emotion* 16, no. 2 (2016): 175–82.

27   **"Post-traumatic growth":** Richard G. Tedeschi and Lawrence G. Calhoun, "Posttraumatic Growth: Conceptual Foundations and Empirical Evidence," *Psychological Inquiry* 15, no. 1 (2004): 1–18.

27   **the great psychologist and Holocaust survivor:** Victor E. Frankl, *Man's Search for Meaning* (New York: Simon and Schuster, 1985; first published in 1946).

27   **make it so:** With apologies to Jean-Luc Picard.

28   **Mindsets affect what people do:** For a wonderful summary, see Carol S. Dweck, *Mindset: The New Psychology of Success* (New York, Random House, 2006). For this particular study, see Ying-yi Hong et al., "Implicit Theories, Attributions, and Coping: A Meaning System Approach," *Journal of Personality and Social Psychology* 77, no. 3 (1999): 588–99.

29   **slightly—but reliably—higher GPAs:** David S. Yeager et al., "Where and for Whom Can a Brief, Scalable Mindset Intervention Improve Adolescents' Educational Trajectories?" (under revision); and Michael Broda et al., "Reducing Inequality in Academic Success for Incoming College Students: A Randomized Trial of Growth Mindset and Belonging Interventions," *Journal of Research on Educational Effectiveness* 11, no. 3 (2018): 317–38.

30   **Carol, Karina, and I:** Interestingly, people's lay theories of empathy didn't always correlate with their overall *level* of empathy. In other

words, people who believe they are empathic don't necessarily think empathy is more under their control. See Karina Schumann et al., "Addressing the Empathy Deficit: Beliefs About the Malleability of Empathy Predict Effortful Responses When Empathy Is Challenging," *Journal of Personality and Social Psychology* 107, no. 3 (2014): 475–93.

CHAPTER 2: CHOOSING EMPATHY

35   *approach motives:* Here I'm using not Lewin's exact language, but the related (and simpler) terminology of Dale Miller and Deborah Prentice, "Psychological Levers of Behavior Change," in *Behavioral Foundations of Policy,* ed. E. Shafir (New York: Russell Sage Foundation, 2010).

35   **the process of buying groceries:** Image from Kurt Lewin, "Group Decision and Social Change," in *Readings in Social Psychology,* ed. Guy Swanson et al. (New York: Henry Holt, 1952), 459–73.

36   **the psychologist William McDougall:** William McDougall, *An Introduction to Social Psychology* (New York: Dover, 2003; first published in 1908).

36   **Many people today agree with him:** Jennifer C. Veilleux et al., "Multidimensional Assessment of Beliefs About Emotion: Development and Validation of the Emotion and Regulation Beliefs Scale," *Assessment* 22, no. 1 (2015): 86–100. Numbers here reflect the proportion of participants who either mildly or strongly agreed with the quoted statements, according to Veilleux, personal communication, July 10, 2018.

37   **"passion is nevertheless weak":** Immanuel Kant, *Groundwork of the Metaphysics of Morals* (New Haven, Conn.: Yale University Press, 2002; first published in 1785.)

37   **this is empathy's fatal flaw:** Paul Bloom, *Against Empathy: The Case for Rational Compassion* (New York: Random House, 2016), 95; and Paul Bloom, "The Baby in the Well: The Case Against Empathy," *New Yorker,* May 20, 2013.

37 **Emotions are *built* on thought:** For more on this perspective, see Lisa Feldman Barrett, *How Emotions Are Made* (New York: Macmillan, 2017); Magda B. Arnold, *Emotion and Personality* (New York: Columbia University Press, 1960); Richard S. Lazarus and Susan Folkman, *Stress, Appraisal, and Coping* (New York: Springer Publishing, 1984); and Klaus R. Scherer et al., *Appraisal Processes in Emotion: Theory, Methods, Research* (Oxford: Oxford University Press, 2001).

38 **people in Gross's studies:** For review, see James J. Gross, "Emotion Regulation: Current Status and Future Prospects," *Psychological Inquiry* 26, no. 1 (2015): 1–26; and Kevin N. Ochsner and James J. Gross, "The Cognitive Control of Emotion," *Trends in Cognitive Sciences* 9, no. 5 (2005): 242–49. Here I'm focusing on rethinking, or, as Gross calls it, "reappraisal." But people also control their feelings in other ways, for instance, by avoiding emotional circumstances or distracting themselves from what they feel.

38 **Tuning helps us constantly:** Lian Bloch et al., "Emotion Regulation Predicts Marital Satisfaction: More Than a Wives' Tale," *Emotion* 14, no. 1 (2014): 130; and Eran Halperin et al., "Can Emotion Regulation Change Political Attitudes in Intractable Conflicts? From the Laboratory to the Field," *Psychological Science* 24, no. 1 (2013): 106–11. Along the lines of Carol Dweck's work, even *believing* that we can control our emotions helps us to do so. For instance, adolescents who believe they can change how they feel are less likely to develop depression. See Brett Q. Ford et al., "The Cost of Believing Emotions Are Uncontrollable," *Journal of Experimental Psychology: General* 147, no. 8 (2018): 1170–90.

38 **people gravitate toward handy emotions:** Maya Tamir, "Why Do People Regulate Their Emotions? A Taxonomy of Motives in Emotion Regulation," *Personality and Social Psychology Review* 20, no. 3 (2016): 199–222.

38 **it *can* occur automatically:** People take on each other's posture, facial expressions, moods, and arousal rapidly, sometimes without even realizing they're doing so. See Celia Heyes, "Automatic Imitation," *Psychological Bulletin* 137, no. 3 (2011): 463–83; Ulf Dimberg et al., "Unconscious Facial Reactions to Emotional Facial Expressions,"

Psychological Science 11, no. 1 (2000): 86–89; and Roland Neumann and Fritz Strack, "'Mood Contagion': The Automatic Transfer of Mood Between Persons," *Journal of Personality and Social Psychology* 79, no. 2 (2000): 211–23.

38   **we choose or avoid it:** For much more of my perspective on this, see Jamil Zaki, "Empathy: A Motivated Account," *Psychological Bulletin* 140, no. 6 (2014): 1608–47.

38   **positive emotion is contagious:** Sylvia A. Morelli et al., "The Emerging Study of Positive Empathy," *Social and Personality Psychology Compass* 9, no. 2 (2015): 57–68.

38   **when people want to connect:** William Ickes et al., "Naturalistic Social Cognition: Empathic Accuracy in Mixed-Sex Dyads," *Journal of Personality and Social Psychology* 59 (1990): 730–42; and Sara Snodgrass, "Women's Intuition: The Effect of Subordinate Role on Interpersonal Sensitivity," *Journal of Personality and Social Psychology* 49, no. 1 (1985): 146–55.

39   **Individuals are more generous in public:** William T. Harbaugh, "The Prestige Motive for Making Charitable Transfers," *American Economic Review* 88, no. 2 (1998): 277–82.

39   **"moral threat":** Eddie Harmon-Jones et al., "The Dissonance-Inducing Effects of an Inconsistency Between Experienced Empathy and Knowledge of Past Failures to Help: Support for the Action-Based Model of Dissonance," *Basic and Applied Social Psychology* 25, no. 1 (2003): 69–78; and Sonya Sachdeva et al., "Sinning Saints and Saintly Sinners: The Paradox of Moral Self-Regulation," *Psychological Science* 20, no. 4 (2009): 523–28.

39   **risks our own well-being:** See, for instance, Sara Hodges and Kristi Klein, "Regulating the Costs of Empathy: The Price of Being Human," *Journal of Socio-Economics* 30, no. 5 (2001): 437–52.

39   **students walked in a wider arc:** S. Mark Pancer et al., "Conflict and Avoidance in the Helping Situation," *Journal of Personality and Social Psychology* 37, no. 8 (1979): 1406–11.

39   **People often avoid empathy:** Laura Shaw et al., "Empathy Avoidance: Forestalling Feeling for Another in Order to Escape the Moti-

vational Consequences," *Journal of Personality and Social Psychology* 67, no. 5 (1994): 879–87.

This logic can also explain compassion collapse. In one set of studies, psychologists presented people with images of either one or eight suffering children. People reported feeling more empathy for one victim than many, but this was especially true of people who excelled at psychological tuning. Worse, training people to tune their emotions made them *less* likely to care. In the face of great suffering, people anticipate that caring will hurt too much. Compassion collapses—not because it's impossible to feel for many victims, but because people *choose* not to. See C. Daryl Cameron and B. Keith Payne, "Escaping Affect: How Motivated Emotion Regulation Creates Insensitivity to Mass Suffering," *Journal of Personality and Social Psychology* 100, no. 1 (2011): 1–15.

40   **Princeton seminary students:** John M. Darley and C. Daniel Batson, "From Jerusalem to Jericho: A Study of Situational and Dispositional Variables in Helping Behavior," *Journal of Personality and Social Psychology* 27, no. 1 (1973): 100–108.

40   **also help themselves:** Morelli, "Emerging Study of Positive Empathy"; Sylvia A. Morelli et al., "Empathy and Well-Being Correlate with Centrality in Different Social Networks," *Proceedings of the National Academy of Sciences* 114, no. 37 (2017): 9843–47; and John F. Helliwell and Lara B. Aknin, "Expanding the Social Science of Happiness," *Nature Human Behaviour* 2 (2018): 248–52.

41   **Individuals who were lonely:** John T. Cacioppo et al., "Reciprocal Influences Between Loneliness and Self-Centeredness: A Cross-Lagged Panel Analysis in a Population-Based Sample of African American, Hispanic, and Caucasian Adults," *Personality and Social Psychology Bulletin* 43, no. 8 (2017): 1125–35.

41   **One of his first cases:** Bernard Burnes, "Kurt Lewin and the Harwood Studies: The Foundations of OD," *Journal of Applied Behavioral Science* 43, no. 2 (2007): 213–31.

42   **Some of these techniques are called "nudges":** This term was coined by Richard H. Thaler and Cass Sunstein in *Nudge: Improving*

Decisions About Health, Wealth, and Happiness (New York, Penguin, 2008). **Retirement saving:** John Beshears et al., "The Importance of Default Options for Retirement Saving Outcomes," in *Social Security Policy in a Changing Environment,* ed. Jeffrey Liebman et al. (Chicago: University of Chicago Press, 2009); **organ donation:** Eric J. Johnson and Daniel Goldstein, "Do Defaults Save Lives?," *Science* 302, no. 5649 (2003): 1338–39.

Nudges are not the only game in town for behavior change; so-called wise interventions take a more explicitly psychological perspective. See, for instance, Gregory M. Walton and Timothy D. Wilson, "Wise Interventions: Psychological Remedies for Social and Personal Problems," *Psychological Review* 125 (2018): 617–55.

43   **Batson flipped compassion collapse:** C. Daniel Batson et al., "Empathy and Attitudes: Can Feeling for a Member of a Stigmatized Group Improve Feelings Toward the Group?," *Journal of Personality and Social Psychology* 72, no. 1 (1997): 105–18.

44   **women *do* exhibit more empathy:** Leonardo Christov-Moore et al., "Empathy: Gender Effects in Brain and Behavior," *Neuroscience and Biobehavioral Reviews* 46 (2014): 604–27.

44   **eliminated the empathic gender gap:** Kristi Klein and Sara Hodges, "Gender Differences, Motivation, and Empathic Accuracy: When It Pays to Understand," *Personality and Social Psychological Bulletin* 27, no. 6 (2001): 720–30; and Geoff Thomas and Gregory R. Maio, "Man, I Feel Like a Woman: When and How Gender-Role Motivation Helps Mind-Reading," *Journal of Personality and Social Psychology* 95, no. 5 (2008): 1165–79.

44   **some of our "selves" are more inclusive:** Samuel L. Gaertner and John F. Dovidio, *Reducing Intergroup Bias: The Common Ingroup Identity Model* (New York: Routledge, 2000).

45   **fans of Manchester United:** Mark Levine et al., "Identity and Emergency Intervention: How Social Group Membership and Inclusiveness of Group Boundaries Shape Helping Behavior," *Personality and Social Psychology Bulletin* 31, no. 4 (2005): 443–53. Note that this study has a relatively small sample size and has not been directly replicated.

But a number of other studies demonstrate that people who focus on a common ingroup do become more prosocial toward outsiders in other contexts.

45   **more likely to be killed by the state:** John F. Edens et al., "Psychopathy and the Death Penalty: Can the Psychopathy Checklist-Revised Identify Offenders Who Represent 'A Continuing Threat to Society'?," *Journal of Psychiatry and Law* 29, no. 4 (2001): 433–81.

46   **Making psychopaths care:** Harma Meffert et al., "Reduced Spontaneous but Relatively Normal Deliberate Vicarious Representations in Psychopathy," *Brain* 136, no. 8 (2013): 2550–62.

47   **put brain mirroring on the map:** Tania Singer et al., "Empathy for Pain Involves the Affective but Not Sensory Components of Pain," *Science* 303, no. 5661 (2004): 1157–62.

47   **test whether these ancient techniques:** Earlier efforts, including from Singer herself, had demonstrated effects of contemplative training on empathy, for instance, Olga M. Klimecki et al., "Differential Pattern of Functional Brain Plasticity After Compassion and Empathy Training," *Social Cognitive and Affective Neuroscience* 9, no. 6 (2014): 873–79; and Paul Condon et al., "Meditation Increases Compassionate Responses to Suffering," *Psychological Science* 24, no. 10 (2013): 2125–27.

48   **What they discovered was striking:** Lea K. Hildebrandt et al., "Differential Effects of Attention-, Compassion-, and Socio-Cognitively Based Mental Practices on Self-Reports of Mindfulness and Compassion," *Mindfulness* 8, no. 6 (2017): 1488–12; Anna-Lena Lumma et al., "Who Am I? Differential Effects of Three Contemplative Mental Trainings on Emotional Word Use in Self-Descriptions," *Self and Identity* 16, no. 5 (2017): 607–28; and Sofie L. Valk et al., "Structural Plasticity of the Social Brain: Differential Change After Socio-Affective and Cognitive Mental Training," *Science Advances* 3, no. 10 (2017): e1700489.

50   **they often convince *themselves*:** For instance, E. Tory Higgins and William S. Rholes, " 'Saying Is Believing': Effects of Message Modification on Memory and Liking for the Person Described," *Journal of Experimental Social Psychology* 14, no. 4 (1978): 363–78.

50 **The exercise stuck:** Erika Weisz et al., "Building Empathy Through Social Psychological Interventions" (in preparation).

CHAPTER 3: HATRED VERSUS CONTACT

53 **Hatred blooms from:** *Race, religion, or gender identity:* Colin Roberts et al., "Understanding Who Commits Hate Crimes and Why They Do It" (report prepared for Welsh Government Social Research, 2013); *unemployment:* Armin Falk et al., "Unemployment and Right-Wing Extremist Crime," *Scandinavian Journal of Economics* 113, no. 2 (2011): 260–85; *history of abuse:* Pete Simi et al., *Trauma as a Precursor to Violent Extremism: How Non-Ideological Factors Can Influence Joining an Extremist Group* (College Park, Md.: START, 2015).

55 **Liberty Net's main message:** Excerpts drawn from transcripts of the lawsuit brought against Tony McAleer by the Canadian Human Rights Council. More can be found at http://www.stopracism.ca/content/chrc-v-canadian-liberty-net1.

55 **whites tacked right politically:** Maureen A. Craig and Jennifer A. Richeson, "On the Precipice of a 'Majority-Minority' America: Perceived Status Threat from the Racial Demographic Shift Affects White Americans' Political Ideology," *Psychological Science* 25, no. 6 (2014): 1189–97; and Robb Willer et al., "Threats to Racial Status Promote Tea Party Support Among White Americans," SSRN working paper, 2016, https://ssrn.com/abstract=2770186.

56 **As our ideals drift further:** See Shanto Iyengar et al., "Affect, Not Ideology: A Social Identity Perspective on Polarization," *Public Opinion Quarterly* 76, no. 3 (2012): 405–31; and Shanto Iyengar and Sean J. Westwood, "Fear and Loathing Across Party Lines: New Evidence on Group Polarization," *American Journal of Political Science* 59, no. 3 (2015): 690–707. Partisanship creeps into our impressions of individuals, too. Iyengar and Westwood asked people to decide between two hypothetical scholarship candidates with comparable records, one of whom belonged to each political party. Both Democrats and Republi-

cans favored candidates from their own party about 80 percent of the time.

56   **Republicans and Democrats paid money:** Jeremy Frimer et al., "Liberals and Conservatives Are Similarly Motivated to Avoid Exposure to One Another's Opinions," *Journal of Experimental Social Psychology* 72 (2017): 1–12.

56   **outsiders in pain:** See, for instance, Xiaojing Xu, "Do You Feel My Pain? Racial Group Membership Modulates Empathic Neural Responses," *Journal of Neuroscience* 29, no. 26 (2009): 8525–29; and John T. Lanzetta and Basil G. Englis, "Expectations of Cooperation and Competition and Their Effects on Observers' Vicarious Emotional Responses," *Journal of Personality and Social Psychology* 56, no. 4 (1989): 543–54.

56   **One physician working in the nineteenth century:** Harriet A. Washington, *Medical Apartheid: The Dark History of Medical Experimentation on Black Americans from Colonial Times to the Present* (New York: Doubleday, 2006).

56   **people guess that a black person:** Sophie Trawalter et al., "Racial Bias in Perceptions of Others' Pain," *PLoS One* 7, no. 11 (2012): e48546; and Kelly M. Hoffmanet et al., "Racial Bias in Pain Assessment and Treatment Recommendations, and False Beliefs About Biological Differences Between Blacks and Whites," *Proceedings of the National Academy of Sciences* 113, no. 16 (2016): 4296–301.

57   **rated Arabs as only about 75 percent evolved:** Nour Kteily et al., "The Ascent of Man: Theoretical and Empirical Evidence for Blatant Dehumanization," *Journal of Personality and Social Psychology* 109, no. 5 (2015): 901–31; and Nour Kteily and Emile Bruneau, "Backlash: The Politics and Real-World Consequences of Minority Group Dehumanization," *Personality and Social Psychology Bulletin* 43, no. 1 (2017): 87–104.

58   **mirroring would be blunted:** This is highlighted by studies using event-related potentials (ERPs), which record brain activity more rapidly than MRIs. See, for instance, Feng Sheng and Shihui Han,

"Manipulations of Cognitive Strategies and Intergroup Relationships Reduce the Racial Bias in Empathic Neural Responses," *Neuroimage* 61, no. 4 (2012): 786–97. ERPs even demonstrate rapid, automatic "counter-empathy" for outsiders within a fraction of a second; see, for instance, Makiko Yamada et al., "Pleasing Frowns, Disappointing Smiles: An ERP Investigation of Counterempathy," *Emotion* 11, no. 6 (2011): 1336.

58   **"schadenfreude," or enjoyment of others' pain: *Brain activity:*** Mina Cikara et al., "Us Versus Them: Social Identity Shapes Neural Responses to Intergroup Competition and Harm," *Psychological Science* 22, no. 3 (2011): 306–13; ***smiling:*** Mina Cikara and Susan T. Fiske, "Stereotypes and Schadenfreude: Affective and Physiological Markers of Pleasure at Outgroup Misfortunes," *Social Psychological and Personality Science* 3, no. 1 (2012): 63–71.

61   **In his magnum opus:** See Gordon W. Allport, *The Nature of Prejudice* (Cambridge, Mass.: Addison Wesley, 1954), especially pp. 269–77; and Thomas F. Pettigrew, "Gordon Willard Allport: A Tribute," *Journal of Social Issues* 55, no. 3 (1999): 415–28.

62   **contact would not always work:** Jens Hainmueller and Daniel J Hopkins, "Public Attitudes Toward Immigration," *Annual Review of Political Science* 17 (2014): 225–49; and Ryan D. Enos, "Causal Effect of Intergroup Contact on Exclusionary Attitudes," *Proceedings of the National Academy of Sciences* 111, no. 10 (2014): 3699–704.

62   **Contact warms sentiments:** Thomas F. Pettigrew and Linda R. Tropp, "A Meta-Analytic Test of Intergroup Contact Theory," *Journal of Personality and Social Psychology* 90, no. 5 (2006): 751. Pettigrew and Tropp find that empathy is one of three main avenues through which contact decreases prejudice; it also makes it harder to maintain stereotypes, and decreases anxiety around outsiders. See Thomas F. Pettigrew and Linda R. Tropp, "How Does Intergroup Contact Reduce Prejudice? Meta-Analytic Tests of Three Mediators," *European Journal of Social Psychology* 38, no. 6 (2008): 922–34.

62   **when people don't seek it out:** Colette Van Laar et al., "The Effect of University Roommate Contact on Ethnic Attitudes and Behavior,"

Journal of Experimental Social Psychology 41, no. 4 (2005): 329–45; and David Broockman and Joshua Kalla, "Durably Reducing Transphobia: A Field Experiment on Door-to-Door Canvassing," *Science* 352, no. 6282 (2016): 220–24.

63 **A linebacker who feels the pain:** For evidence that empathy can interfere with competition, see Debra Gilin et al., "When to Use Your Head and When to Use Your Heart: The Differential Value of Perspective-Taking Versus Empathy in Competitive Interactions," *Personality and Social Psychology Bulletin* 39, no. 1 (2013): 3–16; and Mina Cikara and Elizabeth Levy Paluck, "When Going Along Gets You Nowhere and the Upside of Conflict Behaviors," *Social and Personality Psychology Compass* 7, no. 8 (2013): 559–71.

63 **conservative Israelis reported that:** Roni Porat et al., "What We Want Is What We Get: Group-Based Emotional Preferences and Conflict Resolution," *Journal of Personality and Social Psychology* 110, no. 2 (2016): 167–90.

64 **Contact can build empathy even in the toughest settings:** Tania Tam et al., "The Impact of Intergroup Emotions on Forgiveness in Northern Ireland," *Group Processes and Intergroup Relations* 10, no. 1 (2007): 119–36; Brian M. Johnston and Demis E. Glasford, "Intergroup Contact and Helping: How Quality Contact and Empathy Shape Outgroup Helping," *Group Processes and Intergroup Relations*, July 6, 2017; and Hema Preya Selvanathan et al., "Whites for Racial Justice: How Contact with Black Americans Predicts Support for Collective Action Among White Americans," *Group Processes and Intergroup Relations* 21, no. 6 (2017): 893–912.

64 **the Living Library School Project:** Gábor Orosz et al., "Don't Judge a Living Book by Its Cover: Effectiveness of the Living Library Intervention in Reducing Prejudice Toward Roma and LGBT People," *Journal of Applied Social Psychology* 46, no. 9 (2016): 510–17.

64 **Seeds of Peace:** Juliana Schroeder and Jane L. Risen, "Befriending the Enemy: Outgroup Friendship Longitudinally Predicts Intergroup Attitudes in a Coexistence Program for Israelis and Palestinians," *Group Processes and Intergroup Relations* 19, no. 1 (2016): 72–93.

64   **seventy contact-based programs:** Gunnar Lemmer and Ulrich Wagner, "Can We Really Reduce Ethnic Prejudice Outside the Lab? A Meta-Analysis of Direct and Indirect Contact Interventions," *European Journal of Social Psychology* 45, no. 2 (2015): 152–68.

64   **Allport's rules of engagement:** Though dozens of studies examine contact theory, few truly test Allport's original tenets. See Elizabeth L. Paluck et al., "The Contact Hypothesis Re-Evaluated," *Behavioural Public Policy,* July 10, 2018, 1–30.

65   **where hallucinations originated:** Thomas Dierks et al., "Activation of Heschl's Gyrus During Auditory Hallucinations," *Neuron* 22, no. 3 (1999): 615–21.

67   **the comedian Sarah Silverman:** Drew Magary, "Sarah Silverman Is the Troll Slayer," *GQ Magazine,* May 23, 2018. Consistent with Silverman's perspective, people in low-power positions often do an unusually good job tuning into the minds of those in power. See, for instance, Sara Snodgrass, "Women's Intuition: The Effect of Subordinate Role on Interpersonal Sensitivity," *Journal of Personality and Social Psychology* 49, no. 1 (1985): 146–55; and Michael Kraus et al., "Social Class, Contextualism, and Empathic Accuracy," *Psychological Science* 21, no. 11 (2010): 1716–23.

67   **Emile set up shop:** Emile Bruneau and Rebecca Saxe, "The Power of Being Heard: The Benefits of 'Perspective-Giving' in the Context of Intergroup Conflict," *Journal of Experimental Social Psychology* 48, no. 4 (2012): 855–66.

68   **Today's white nationalists follow suit:** Patrick S. Forscher and Nour Kteily, "A Psychological Profile of the Alt-Right," working paper, 2017, https://psyarxiv.com/c9uvw.

69   **Tony, Emile, Nour Kteily, and I all gathered:** The meeting was convened and spearheaded by Rachel Brown, the executive director of Over Zero, a nonprofit dedicated to reducing and preventing violence. Rachel has engaged in innovative and powerful advocacy against hatred around the world, including in Kenya and the United States.

70   **As she has recalled:** Note that this quote comes not from our conversation, but from an interview Angela participated in: A. Schmidt, "Former Neo-Nazi Who Joined Hate Group at 15 and Changed Her Life Now Helps Other Ex-Racists Leave Violent Extremism," *Daily Mail,* March 7, 2017.

71   **Self-compassion and empathy toward others:** Kristin D. Neff and Elizabeth Pommier, "The Relationship Between Self-Compassion and Other-Focused Concern Among College Undergraduates, Community Adults, and Practicing Meditators," *Self and Identity* 12, no. 2 (2013): 160–76. Neff and Pommier found that correlations between self-compassion and empathic concern ranged from 0 to .26. Even the high end of this range is what statisticians would consider a weak relationship.

71   **People who lack self-compassion:** Lisa M. Yarnell and Kristin D. Neff, "Self-Compassion, Interpersonal Conflict Resolutions, and Well-Being," *Self and Identity* 12, no. 2 (2013): 146–59.

71   **Israeli children trained in self-compassion:** Rony Berger et al., "A School-Based Intervention for Reducing Posttraumatic Symptomatology and Intolerance During Political Violence," *Journal of Educational Psychology* 108, no. 6 (2016): 761–71. Note that this program focused not only on self-compassion but also on stress reduction.

72   **people who can vividly imagine their future self:** Hal Ersner-Hershfield et al., "Don't Stop Thinking About Tomorrow: Individual Differences in Future Self-Continuity Account for Saving," *Judgment and Decision Making* 4, no. 4 (2009): 280–86; and Hal Ersner-Hershfield et al., "Increasing Saving Behavior Through Age-Progressed Renderings of the Future Self," *Journal of Marketing Research* 48 (2011): S23–S37.

73   **convince Israelis and Palestinians:** Amit Goldenberg et al., "Testing the Impact and Durability of a Group Malleability Intervention in the Context of the Israeli–Palestinian Conflict," *Proceedings of the National Academy of Sciences* 115, no. 4 (2018): 696–701; and Amit Goldenberg et al., "Making Intergroup Contact More Fruitful:

Enhancing Cooperation Between Palestinian and Jewish-Israeli Adolescents by Fostering Beliefs About Group Malleability," *Social Psychological and Personality Science* 8, no. 1 (2016).

CHAPTER 4: THE STORIES WE TELL

74   **mental voyages drag your senses along:** Mark K. Wheeler et al., "Memory's Echo: Vivid Remembering Reactivates Sensory-Specific Cortex," *Proceedings of the National Academy of Sciences* 97, no. 20 (2000): 11125–29; Lars Nyberg et al., "Reactivation of Motor Brain Areas During Explicit Memory for Actions," *Neuroimage* 14, no. 2 (2001): 521–28; and Bruno Laeng and Unni Sulutvedt, "The Eye Pupil Adjusts to Imaginary Light," *Psychological Science* 25, no. 1 (2014): 188–97.

75   **They reported being less happy:** Matthew A. Killingsworth and Daniel. T. Gilbert, "A Wandering Mind Is an Unhappy Mind," *Science* 330, no. 6006 (2010): 932.

75   **a set of brain regions:** These regions are commonly referred to as the "default mode network," because their default state is metabolically active, as opposed to inactive. For early writing on this network, see Marcus E. Raichle et al., "A Default Mode of Brain Function," *Proceedings of the National Academy of Sciences* 98, no. 2 (2001): 676–82.

75   **a steering system for untethering:** For a great review of this system's involvement in untethering, see Randy L. Buckner and Daniel C. Carroll, "Self-Projection and the Brain," *Trends in Cognitive Sciences* 11, no. 2 (2007): 49–57.

76   **The more you engage:** Diana I. Tamir and Jason P. Mitchell, "Neural Correlates of Anchoring-and-Adjustment During Mentalizing," *Proceedings of the National Academy of Sciences* 107, no. 24 (2010): 10827–32; and Jamil Zaki et al., "The Neural Bases of Empathic Accuracy," *Proceedings of the National Academy of Sciences* 106, no. 27 (2009): 11382–87.

76   **much of our free time pretending:** For more, see Jonathan Gottschall, *The Storytelling Animal: How Stories Make Us Human* (New York: Houghton Mifflin Harcourt, 2012).

76 **performance-enhancing drugs for untethering:** Narrative arts activate the same brain regions as memory, imagination, and empathy.
This is consistent with the idea that they hack into our experience of
all three. See Raymond Mar, "The Neural Bases of Social Cognition
and Story Comprehension," *Annual Review of Psychology* 62 (2011):
103–34.

78 **"[The actor's] job":** Konstantin Stanislavsky, *An Actor Prepares*,
trans. Elizabeth Reynolds Hapgood (New York: Routledge, 1989; first
published in 1936), 15.

79 **cognitive empathy and theater:** Goldstein used the term "theory
of mind," a term that overlaps heavily with cognitive empathy (see
appendix A, "What Is Empathy?"). I substitute here for the sake of
consistency.

80 **Children with active imaginations outperformed:** M. Taylor and
S. M. Carlson, "The Relation Between Individual Differences in
Fantasy and Theory of Mind," *Child Development* 68, no. 3 (1997):
436–55.

80 **As Goldstein wrote:** Thalia R. Goldstein et al., "Actors Are Skilled
in Theory of Mind but Not Empathy," *Imagination, Cognition and
Personality* 29, no. 2 (2009): 115–33.

80 **theater training also *grew* students' cognitive empathy:** Thalia R.
Goldstein and Ellen Winner, "Enhancing Empathy and Theory of
Mind," *Journal of Cognition and Development* 13, no. 1 (2012): 19–37.
Other forms of art do have well-documented positive effects on other
psychological functions, for instance improving makers' well-being;
see, for instance, Louise C. Boyes and Ivan Reid, "What Are the Ben-
efits for Pupils Participating in Arts Activities? The View from the
Research Literature," *Research in Education* 73, no. 1 (2005): 1–14.

81 **Kids with autism:** Blythe A. Corbett et al., "Improvement in Social
Competence Using a Randomized Trial of a Theatre Intervention for
Children with Autism Spectrum Disorder," *Journal of Autism and
Developmental Disorders* 46, no. 2 (2016): 658–72.

81 **Raymond Mar has examined:** For a great summary of Mar's perspective, see Raymond Mar and Keith Oatley, "The Function of Fiction Is

the Abstraction and Simulation of Social Experience," *Perspectives on Psychological Science* 3, no. 3 (2008): 173–92.

81   **avid readers have an easier time:** For a meta-analysis of correlational studies establishing a fiction-empathy connection, see Micah L. Mumper and Richard J. Gerrig, "Leisure Reading and Social Cognition: A Meta-Analysis," *Psychology of Aesthetics, Creativity, and the Arts* 11, no. 1 (2017): 109–20. Interestingly, the amount of technical nonfiction people read does not track their empathy, so stodgy academics like me are out of luck.

81   **Small "doses" of fiction:** For a meta-analysis of these effects, see David Dodell-Feder and Diana I. Tamir, "Fiction Reading Has a Small Positive Impact on Social Cognition: A Meta-Analysis," *Journal of Experimental Psychology: General* (2018). See also Maria C. Pino and M. Mazza, "The Use of 'Literary Fiction' to Promote Mentalizing Ability," *PLoS One* 11, no. 8 (2016): e0160254; ***grief and giving:*** Eve M. Koopman, "Empathic Reactions After Reading: The Role of Genre, Personal Factors and Affective Responses," *Poetics* 50 (2015): 62–79.

Research on empathy and fiction includes non-replications of a prominent early finding; for instance, Maria E. Panero et al., "Does Reading a Single Passage of Literary Fiction Really Improve Theory of Mind? An Attempt at Replication," *Journal of Personality and Social Psychology* 111, no. 5 (2016): e46–e54. But the weight of evidence points to a consistent, if small, effect. As Dodell-Feder and Tamir point out, even a small effect is meaningful, because most interventions in these studies are quite minimal. If reading one passage of fiction (or one book) creates a tiny increase in empathy, a lifetime of doing so might create a larger difference.

82   **This pattern pops up:** Philip J. Mazzocco et al., "This Story Is Not for Everyone: Transportability and Narrative Persuasion," *Social Psychological and Personality Science* 1, no. 4 (2010): 361–68; Loris Vezzali et al., "Indirect Contact Through Book Reading: Improving Adolescents' Attitudes and Behavioral Intentions Toward Immigrants," *Psychology in the Schools* 49, no. 2 (2012): 148–62; and Dan R. Johnson, "Transportation into Literary Fiction Reduces Prejudice Against and

Increases Empathy for Arab-Muslims," *Scientific Study of Literature* 3, no. 1 (2013): 77–92.

83   **as one survivor described it:** Betsy Levy Paluck, personal communication, August 30, 2016.

83   **Jean Chrétien describes:** As quoted in Charles Mironko, "The Effect of RTLM's Rhetoric of Ethnic Hatred in Rural Rwanda," in *The Media and the Rwandan Genocide*, ed. Allan Thompson (London: Pluto Press, 2007), 125.

84   **It hired charismatic announcers:** Monologues in fact composed about 61 percent of all airtime content on RTLM. See Mary Kimani, "RTLM: The Medium That Became a Tool for Mass Murder," in Thompson, *Media and the Rwandan Genocide*.

84   **"Our priority is to assist all Rwandans":** RTLM transcript, January 14, 1994; all RTLM transcripts quoted here available at http://www.rwandafile.com/rtlm/.

84   **Many Rwandans remember these programs fondly:** Betsy Levy Paluck, personal communication, August 30, 2016: "I always pictured RTLM as menacing, polemic, and threatening, but people described how amusing it was, how funny, and how high-spirited. Someone would recount things that were on that radio station, and people would still laugh."

84   ***gacaca* also publicly reopened psychological wounds:** Cilliers et al., "Reconciling After Civil Conflicts Increases Social Capital but Decreases Individual Well-Being," *Science* 352, no. 6287 (2016): 787–94. They found that following truth and reconciliation programs in Sierra Leone, victims actually experience an increase in PTSD-related symptoms. Although similar data have not been collected from Rwanda, Paluck's experience with post-genocide emotion jibe with Cilliers's and colleagues' account.

85   **Paluck found that *New Dawn*:** Elizabeth Levy Paluck, "Reducing Intergroup Prejudice and Conflict Using the Media: A Field Experiment in Rwanda," *Journal of Personality and Social Psychology* 96, no. 3 (2009): 574–87.

86   **who heard Batamuriza's voice:** Rezarta Bilali and Johanna R. Voll-hardt, "Priming Effects of a Reconciliation Radio Drama on Historical Perspective-Taking in the Aftermath of Mass Violence in Rwanda," *Journal of Experimental Social Psychology* 49, no. 1 (2013): 144–51.

87   **The Bureau of Justice Statistics:** Matthew R. Durose et al., *Recidivism of Prisoners Released in 30 States in 2005: Patterns from 2005 to 2010* (Washington, D.C.: U.S. Department of Justice, Office of Justice Programs, Bureau of Justice Statistics, 2014).

88   **"they're going to steal all our computers":** A couple of weeks into the program, many computers *were* stolen from campus. The administration immediately called Waxler, sure that his students were responsible. But soon after, the actual thief—a "regular" UMass student—was caught.

91   **less than 20 percent of Changing Lives students:** G. Roger Jarjoura and Susan T. Krumholz, "Combining Bibliotherapy and Positive Role Modeling as an Alternative to Incarceration," *Journal of Offender Rehabilitation* 28, nos. 1–2 (1998): 127–39; *follow-up study:* Russell K. Schutt et al., "Using Bibliotherapy to Enhance Probation and Reduce Recidivism," *Journal of Offender Rehabilitation* 52, no. 3 (2013): 181–97.

92   **28 percent of all New York City schools:** New York City Comptroller's Office, *State of the Arts: A Plan to Boost Arts Education in New York City Schools,* April 7, 2014.

CHAPTER 5: CARING TOO MUCH

96   **No single emotional experience:** *Anxiety:* Alia J. Crum et al., "Rethinking Stress: The Role of Mindsets in Determining the Stress Response," *Journal of Personality and Social Psychology* 104, no. 4 (2013): 716; *joy:* June Gruber et al., "A Dark Side of Happiness? How, When, and Why Happiness Is Not Always Good," *Perspectives on Psychological Science* 6, no. 3 (2011): 222–33.

96   **"compassion fatigue":** Carla Joinson, "Coping with Compassion Fatigue," *Nursing* 22, no. 4 (1992): 116–18; and Charles R. Figley,

"Compassion Fatigue: Psychotherapists' Chronic Lack of Self Care," *Journal of Clinical Psychology* 58, no. 11 (2002): 1433–41.

96 **Caregivers are much more likely:** Karlijn J. Joling et al., "Incidence of Depression and Anxiety in the Spouses of Patients with Dementia: A Naturalistic Cohort Study of Recorded Morbidity with a 6-Year Follow-Up," *American Journal of Geriatric Psychiatry* 18, no. 2 (2010): 146–53; and Martin Pinquart and Silvia Sorensen, "Differences Between Caregivers and Noncaregivers in Psychological Health and Physical Health: A Meta-Analysis," *Psychology and Aging* 18, no. 2 (2003): 250–67.

96 **empathic parents paid a price:** Erika Manczak et al., "Does Empathy Have a Cost? Diverging Psychological and Physiological Effects Within Families," *Health Psychology* 35, no. 3 (2016): 211; and Erika Manczak et al., "The Price of Perspective Taking: Child Depressive Symptoms Interact with Parental Empathy to Predict Immune Functioning in Parents," *Clinical Psychological Science* 4, no. 3 (2017): 485–92.

97 **"news fatigue":** Katherine N. Kinnick et al., "Compassion Fatigue: Communication and Burnout Toward Social Problems," *Journalism and Mass Communication Quarterly* 73, no. 3 (1996): 687–707; and Jeffrey Gottfried and Michael Barthel, "Almost Seven-in-Ten Americans Have News Fatigue, More Among Republicans," Pew Research Center, June 5, 2018.

98 **The first baby we see:** To protect the privacy of patients and their families, I've altered the details of their cases, though they are representative of babies on the unit during my visit. This is true of Francisco's case in particular.

99 **They startle no one:** "Alarm fatigue" is prevalent among intensive care units; doctors and nurses become so accustomed to warning sounds that they begin to ignore them, endangering patients. See, for instance, Sue Sendelbach and Marjorie Funk, "Alarm Fatigue: A Patient Safety Concern," *AACN Advanced Critical Care* 24, no. 4 (2013): 378–86.

102 **"understanding binds, but feeling blinds":** Mohammadreza Hojat, *Empathy in Health Professions Education and Patient Care* (New

York: Springer, 2016). For instance, Hojat writes, "Emotional empathy in excess . . . can be detrimental to objectivity in clinical decision-making." Ibid., 80. For more on this perspective, as well as some challenges to it, see Jodi Halpern, *From Detached Concern to Empathy: Humanizing Medical Practice* (Oxford: Oxford University Press, 2001).

102 **"physicians' feelings are extraneous":** Diane E. Meier et al., "The Inner Life of Physicians and Care of the Seriously Ill," *JAMA* 286, no. 23 (2001): 3007–14.

103 **sugarcoat bad news:** See, for instance, Carol F. Quirt et al., "Do Doctors Know When Their Patients Don't? A Survey of Doctor-Patient Communication in Lung Cancer," *Lung Cancer* 18, no. 1 (1997): 1–20; and Lesley Fallowfield and Val A. Jenkins, "Communicating Sad, Bad, and Difficult News in Medicine," *Lancet* 363, no. 9405 (2004): 312–19.

103 **"secondary trauma":** Meredith Mealer et al., "Increased Prevalence of Post-Traumatic Stress Disorder Symptoms in Critical Care Nurses," *American Journal of Respiratory and Critical Care Medicine* 175, no. 7 (2007): 693–97.

103 **One in three intensive care professionals:** Aynur Aytekin et al., "Burnout Levels in Neonatal Intensive Care Nurses and Its Effects on Their Quality of Life," *Australian Journal of Advanced Nursing* 31, no. 2 (2013): 39; Nathalie Embriaco et al., "High Level of Burnout in Intensivists: Prevalence and Associated Factors," *American Journal of Respiratory and Critical Care Medicine* 175, no. 7 (2007): 686–92; and Margot M. C. van Mol et al., "The Prevalence of Compassion Fatigue and Burnout Among Healthcare Professionals in Intensive Care Units: A Systematic Review," *PLoS One* 10, no. 8 (2015): e0136955.

103 **Empathic professionals bear the brunt:** Kevin J. Corcoran, "Interpersonal Stress and Burnout: Unraveling the Role of Empathy," *Journal of Social Behavior and Personality* 4, no. 1 (1989): 141–44; Carol A. Williams, "Empathy and Burnout in Male and Female Helping Professionals," *Research in Nursing and Health* 12, no. 3 (1989): 169–78; and Colin P. West et al., "Association of Perceived Medical Errors

with Resident Distress and Empathy: A Prospective Longitudinal Study," *JAMA* 296, no. 9 (2006): 1071–78.

104   **At the outset of their training:** *Medical students:* Matthew R. Thomas et al., "How Do Distress and Well-Being Relate to Medical Student Empathy? A Multicenter Study," *Journal of General Internal Medicine* 22, no. 2 (2007): 177–83; *nursing students:* Paula Nunes et al., "A Study of Empathy Decline in Students from Five Health Disciplines During Their First Year of Training," *International Journal of Medical Education* 2 (2011): 12–17.

104   **Patients of empathic physicians:** Mohammadreza Hojat et al., "Physicians' Empathy and Clinical Outcomes for Diabetic Patients," *Academic Medicine* 86, no. 3 (2011): 359–64; and Sung Soo Kim et al., "The Effects of Physician Empathy on Patient Satisfaction and Compliance," *Evaluation and the Health Professions* 27, no. 3 (2004): 237–51.

104   **burning out, quitting, or both:** Social workers can experience astonishing rates of turnover—in some settings, between 30 to 60 percent leave their jobs each year. Michàl. E. Mor Barak et al., "Antecedents to Retention and Turnover Among Child Welfare, Social Work, and Other Human Service Employees: A Review and Metanalysis," *Social Service Review* 75, no. 4 (2001): 625–61; see also Mercedes Braithwaite, "Nurse Burnout and Stress in the NICU," *Advances in Neonatal Care* 8, no. 6 (2008): 343–47.

104   **by their third year, they empathize** *less***:** Melanie Neumann et al., "Empathy Decline and Its Reasons: A Systematic Review of Studies with Medical Students and Residents," *Academic Medicine* 86, no. 8 (2011): 996–1009; and Mohammadreza Hojat et al., "The Devil Is in the Third Year: A Longitudinal Study of Erosion of Empathy in Medical School," *Academic Medicine* 84, no. 9 (2009): 1182–91.

104   **underestimate patients' suffering:** Rod Sloman et al., "Nurses' Assessment of Pain in Surgical Patients," *Journal of Advanced Nursing* 52, no. 2 (2005): 125–32; Lisa J. Staton et al., "When Race Matters: Disagreement in Pain Perception Between Patients and Their Physicians in Primary Care," *Journal of the National Medical Association* 99, no. 5 (2007): 532. Medical professionals also exhibit blunted brain

mirroring of pain; see, for instance, Jean Decety et al., "Physicians Down-Regulate Their Pain Empathy Response: An Event-Related Brain Potential Study," *Neuroimage* 50, no. 4 (2010): 1676–82.

104   **dehumanized patients:** Omar S. Haque and Adam Waytz, "Dehumanization in Medicine Causes, Solutions, and Functions," *Perspectives on Psychological Science* 7, no. 2 (2012): 176–86; and Jereon Vaes and Martina Muratore, "Defensive Dehumanization in the Medical Practice: A Cross-Sectional Study from a Health Care Worker's Perspective," *British Journal of Social Psychology* 52, no. 1 (2013): 180–90.

105   **less than fifteen minutes per visit:** *Conversation time:* See Greg Irving et al., "International Variations in Primary Care Physician Consultation Time: A Systematic Review of 67 Countries," *BMJ Open* 7, no. 10 (2017): e017902; and Christine Sinsky et al., "Allocation of Physician Time in Ambulatory Practice: A Time and Motion Study in 4 Specialties," *Annals of Internal Medicine* 165, no. 11 (2016): 753–60. Irving et al. report an average visit length of about twenty minutes in the United States, and Sinsky et al. estimate that about 50 percent of each visit is spent in conversation; *interruptions:* M. Kim Marvel et al., "Soliciting the Patient's Agenda: Have We Improved?" *JAMA* 281, no. 3 (1999): 283–87.

105   **physician burnout crept upward:** Tait D. Shanafelt et al., *Changes in Burnout and Satisfaction with Work-Life Balance in Physicians and the General U.S. Working Population Between 2011 and 2014* (paper presented at the Mayo Clinic Proceedings, 2015).

105   **as marathon twenty-four-hour shifts:** Stacey A. Passalacqua and Chris Segrin, "The Effect of Resident Physician Stress, Burnout, and Empathy on Patient-Centered Communication During the Long-Call Shift," *Health Communication* 27, no. 5 (2012): 449–56.

106   **protects caregivers from burnout:** Michael Kearney et al., "Self-Care of Physicians Caring for Patients at the End of Life: 'Being Connected . . . a Key to My Survival,'" *JAMA* 301, no. 11 (2009): 1155–64; and Sandra Sanchez-Reilly et al., "Caring for Oneself to Care for Others: Physicians and Their Self-Care," *Journal of Supportive Oncology* 11, no. 2 (2013): 75–81.

106 **therapists who used self-care:** Ted Bober and Cheryl Regehr, "Strategies for Reducing Secondary or Vicarious Trauma: Do They Work?" *Brief Treatment and Crisis Intervention* 6, no. 1 (2006): 1–9.

106 **Social support buffers nurses and doctors:** Richard Jenkins and Peter Elliott, "Stressors, Burnout and Social Support: Nurses in Acute Mental Health Settings," *Journal of Advanced Nursing* 48, no. 6 (2004): 622–31.

108 **Josie King:** For more on Josie's story, see Erika Niedowski, "How Medical Errors Took a Little Girl's Life," *Baltimore Sun,* December 14, 2003.

108 **Hopkins overhauled their practices:** For instance, they automated their prescription pipeline to minimize possible human error. They also instituted clear checklists for common procedures such as inserting IVs, where inattention to little details can cause infection, and empowered nurses and other staff to question physicians whenever they missed a step.

109 **Wu interviewed residents:** Albert W. Wu, "Medical Error: The Second Victim," *BMJ: British Medical Journal* 320, no. 7237 (2000): 726.

109 **After residents made an error:** West, "Association of Perceived Medical Errors with Resident Distress and Empathy."

109 **"psychological first aid":** The evidence on psychological first aid for disaster-related trauma is mixed, but at least some research supports the idea that "short-circuiting" responses to trauma can prevent or mitigate PTSD. For instance, when trauma victims take propranolol (a beta blocker) while remembering traumatic events, the typical surge of stress hormones they experience is diminished, and so is their subsequent PTSD. See Roger K. Pitman et al., "Pilot Study of Secondary Prevention of Posttraumatic Stress Disorder with Propranolol," *Biological Psychiatry* 51, no. 2 (2002): 189–92.

110 **Wu and his team found:** Hanan Edrees et al., "Implementing the RISE Second Victim Support Programme at the Johns Hopkins Hospital: A Case Study," *BMJ Open* 6, no. 9 (2016): e011708.

112 **"unexamined emotions":** Meier, "Inner Life of Physicians and Care of the Seriously Ill."

112   **"emotional granularity":** For a recent account of this phenomenon, see Katharine E. Smidt and Michael K. Suvak, "A Brief, but Nuanced, Review of Emotional Granularity and Emotion Differentiation Research," *Current Opinion in Psychology* 3 (2015): 48–51.

112   **People who could pinpoint their feelings:** Lisa Feldman Barrett et al., "Knowing What You're Feeling and Knowing What to Do About It: Mapping the Relation Between Emotion Differentiation and Emotion Regulation," *Cognition and Emotion* 15, no. 6 (2001): 713–24.

112   **High-granularity individuals:** Todd B. Kashdan et al., "Unpacking Emotion Differentiation: Transforming Unpleasant Experience by Perceiving Distinctions in Negativity," *Current Directions in Psychological Science* 24, no. 1 (2015): 10–16; and Landon F. Zaki et al., "Emotion Differentiation as a Protective Factor Against Nonsuicidal Self-Injury in Borderline Personality Disorder," *Behavior Therapy* 44, no. 3 (2013): 529–40.

113   **One program taught schoolchildren:** Mark A. Brackett et al., "Enhancing Academic Performance and Social and Emotional Competence with the RULER Feeling Words Curriculum," *Learning and Individual Differences* 22, no. 2 (2012): 218–24.

114   **They are only weakly related:** Mark Davis, "Measuring Individual Differences in Empathy: Evidence for a Multidimensional Approach," *Journal of Personality and Social Psychology* 44, no. 1 (1983): 113–26; and Matthew R. Jordan et al., "Are Empathy and Concern Psychologically Distinct?" *Emotion* 16, no. 8 (2016): 1107–16.

114   **also inspire different actions:** Mark Davis et al., "Empathy, Expectations, and Situational Preferences: Personality Influences on the Decision to Participate in Volunteer Helping Behaviors," *Journal of Personality* 67, no. 3 (1999): 469–503; and C. Daniel Batson and Laura L. Shaw, "Evidence for Altruism: Toward a Pluralism of Prosocial Motives," *Psychological Inquiry* 2, no. 2 (1991): 107–22.

114   **only distress tracks burnout:** See, for instance, Ezequiel Gleichgerrcht and Jean Decety, "Empathy in Clinical Practice: How In-

dividual Dispositions, Gender, and Experience Moderate Empathic Concern, Burnout, and Emotional Distress in Physicians," *PLoS One* 8, no. 4 (2013): e61526; and Martin Lamothe et al., "To Be or Not to Be Empathic: The Combined Role of Empathic Concern and Perspective Taking in Understanding Burnout in General Practice," *BMC Family Practice* 15, no. 1 (2014): 15–29.

115 **Having spent time with her:** During one of our interviews, I administered the Interpersonal Reactivity Index, a popular empathy questionnaire, to Liz. She scored off the charts on concern but seemed almost totally immune to distress.

115 **people who practiced *metta*:** Olga M. Klimecki et al., "Differential Pattern of Functional Brain Plasticity After Compassion and Empathy Training," *Social Cognitive and Affective Neuroscience* 9, no. 6 (2014): 873–79.

116 **physicians who completed meditation-based programs:** For a review and meta-analysis on this type of program, see Colin P. West et al., "Interventions to Prevent and Reduce Physician Burnout: A Systematic Review and Meta-Analysis," *Lancet* 388, no. 10057 (2016): 2272–81.

116 **Eve is connecting these dots:** For more on her persective, see Eve Ekman and Michael Krasner, "Empathy in Medicine: Neuroscience, Education and Challenges," *Medical Teacher* 39, no. 2 (2017): 164–73; and Eve Ekman and Jodi Halpern, "Professional Distress and Meaning in Health Care: Why Professional Empathy Can Help," *Social Work in Health Care* 54, no. 7 (2015): 633–50.

116 **emerging evidence is promising:** Jennifer S. Mascaro et al., "Meditation Buffers Medical Student Compassion from the Deleterious Effects of Depression," *Journal of Positive Psychology* 13, no. 2 (2018): 133–42.

117 **survival rates for extremely premature babies:** Noelle Young et al., "Survival and Neurodevelopmental Outcomes Among Periviable Infants," *New England Journal of Medicine* 376, no. 7 (2017): 617–28.

117   **caregivers who think their job:** Anthony L. Back et al., " 'Why Are We Doing This?': Clinician Helplessness in the Face of Suffering," *Journal of Palliative Medicine* 18, no. 1 (2015): 26–30.

CHAPTER 6: KIND SYSTEMS

119   **Norms affect us:** *Evidence about norms and food:* Erik C. Nook and Jamil Zaki, "Social Norms Shift Behavioral and Neural Responses to Foods," *Journal of Cognitive Neuroscience* 27, no. 7 (2015): 1412–26; *attractiveness:* Jamil Zaki et al., "Social Influence Modulates the Neural Computation of Value," *Psychological Science* 22, no. 7 (2011): 894–900; *voting:* Robert M. Bond et al., "A 61-Million-Person Experiment in Social Influence and Political Mobilization," *Nature* 489, no. 7415 (2012): 295–98; *emotions:* Amit Goldenberg et al., "The Process Model of Group-Based Emotion: Integrating Intergroup Emotion and Emotion Regulation Perspectives," *Personality and Social Psychology Review* 20, no. 2 (2016): 118–41.

119   **Psychologists once interviewed freshmen:** Deborah A. Prentice and Dale T. Miller, "Pluralistic Ignorance and Alcohol Use on Campus: Some Consequences of Misperceiving the Social Norm," *Journal of Personality and Social Psychology* 64, no. 2 (1993): 243.

120   **extreme voices on cable news:** A recent survey found that the United States has one of the most polarized media ecosystems in the Western world; see Brett Etkins, "U.S. Media Among Most Polarized in the World," *Forbes,* June 27, 2017. Consuming these extreme opinions tends to leave viewers more partisan and less tolerant of outsiders as well. See, for instance, Matthew Levendusky, "Partisan Media Exposure and Attitudes Toward the Opposition," *Political Communication* 30, no. 4 (2013): 565–81.

121   **moral revolutions begin:** Kwame Anthony Appiah, *The Honor Code: How Moral Revolutions Happen* (New York: W. W. Norton, 2010).

121   **We catch one another's empathy:** Erik C. Nook et al., "Prosocial Conformity: Generalization Across Behavior and Affect," *Personality and Social Psychology Bulletin* 42, no. 8 (2016): 1045–62.

122 **reduced students' alcohol use:** Christine M. Schroeder and Deborah A. Prentice, "Exposing Pluralistic Ignorance to Reduce Alcohol Use Among College Students," *Journal of Applied Social Psychology* 28, no. 23 (1998): 2150–80.

122 **Google found that its most successful teams:** Charles Duhhig, "What Google Learned from Its Quest to Build the Perfect Team," *New York Times,* February 25, 2016; Anita W. Woolley et al., "Evidence for a Collective Intelligence Factor in the Performance of Human Groups," *Science* 330, no. 6004 (2010): 686–88; and Phillip M. Podsakoff and Scott B. MacKenzie, "Impact of Organizational Citizenship Behavior on Organizational Performance: A Review and Suggestion for Future Research," *Human Performance* 10, no. 2 (1997): 133–51.

122 **IDEO encourages employees:** Teresa Amabile et al., "IDEO's Culture of Helping," *Harvard Business Review,* January–February 2014.

123 **Modern policing is a surprisingly young line of work:** For much more on the history of policing, from Peel to the warrior mentality, see Seth W. Stoughton, "Principled Policing: Warrior Cops and Guardian Officers," *Wake Forest Law Review* 51, 611 (2016).

124 **They were rewarded:** This was, of course, not an unmitigated good. "Political policing," as Stoughton refers to it, produced a great deal of corruption and inside dealing—with officers letting community insiders slide for common crimes—and violence toward outsiders. See Stoughton, "Principled Policing."

125 **In 2014, course materials:** Uriel J. Garcia, "Experts Say Strongly Worded Police Curriculum Is Risky with Cadets," *Santa Fe New Mexican,* March 22, 2014.

125 **Grossman delivers "The Bulletproof Warrior":** For more on Grossman, see Radley Balko, "A Day with 'Killology' Police Trainer Dave Grossman," *Washington Post,* February 14, 2017. For evidence on the number of police killed in the line of duty, see "US Police Shootings: How Many Die Each Year?" *BBC Magazine,* July 18, 2016; and the FBI's Uniform Crime Reporting Project, at https://www.fbi.gov /services/cjis/ucr/publications#LEOKA. For firearm usage data, see

Rich Morin and Andrew Mercer, "A Closer Look at Police Officers Who Have Fired Their Weapon on Duty," PewResearch.org, February 8, 2017.

126   **also make violence more likely:** For more on the weapons task, see B. Keith Payne, "Weapon Bias: Split-Second Decisions and Unintended Stereotyping," *Current Directions in Psychological Science* 15, no. 6 (2006): 287–91. For evidence that stress worsens weapons bias, see Arne Nieuwenhuys et al., "Shoot or Don't Shoot? Why Police Officers Are More Inclined to Shoot When They Are Anxious," *Emotion* 12, no. 4 (2012): 827–33.

127   **almost five civilians were killed by police officers per day:** It is surprisingly difficult to calculate the exact number of people killed in encounters with police. Records are vastly decentralized, and kept by police departments with a vested interest in minimizing the appearance of abuse or bias. One citizen-led project that has kept strong records is Brian Burghart's Fatal Encounters, which can be found at http://www.fatalencounters.org/.

127   **a two-decade low point:** See "In U.S., Confidence in Police Lowest in 22 Years," Gallup, June 2015, and "Race Relations," Gallup, April 2018.

128   **Rahr's most important mandate:** See Sue Rahr and Stephen K. Rice, *From Warriors to Guardians: Recommitting American Police Culture to Democratic Ideals,* U.S. Department of Justice, Office of Justice Programs, National Institute of Justice, 2015.

128   **Tom Tyler has demonstrated:** Tom R. Tyler and E. Allan Lind, "A Relational Model of Authority in Groups," *Advances in Experimental Social Psychology* 25 (1992): 115–91.

131   **officers used force 30 percent less often:** See Jacqueline Helfgott et al., "The Effect of Guardian Focused Training for Law Enforcement Officers," Seattle University Department of Criminal Justice, 2017; and Emily Owens et al., "Promoting Officer Integrity Through Early Engagements and Procedural Justice in the Seattle Police Department," report submitted to the Department of Justice, project no. 2012-IJ-CX-0009, 2016.

134 **look like executions:** Nonlethal interactions don't instill confidence, either. My colleague Jennifer Eberhardt recently analyzed data from more than twenty-eight thousand traffic stops made by Oakland police officers between 2013 and 2014. Police officers were more likely to handcuff, search, and arrest black versus white citizens, even when controlling for neighborhood crime rates and numerous other factors. Using audio from body-worn cameras, Eberhardt and her team could determine the race of a citizen based only on the language officers used, for instance, discussing probation more often with black citizens, even those who had committed no crimes. See Rob Voigt et al., "Language from Police Body Camera Footage Shows Racial Disparities in Officer Respect," *Proceedings of the National Academy of Sciences* 114, no. 25 (2017): 6521–56.

134 **cops often circle the wagons:** Rich Morin et al., "Police, Fatal Encounters, and Ensuing Protests," Pew Research Center, January 11, 2017.

134 **"empathy bias":** Emile G. Bruneau et al., "Parochial Empathy Predicts Reduced Altruism and the Endorsement of Passive Harm," *Social Psychological and Personality Science* 8, no. 8 (2017): 934–42.

137 **In schools, these policies are meant:** Russell J. Skiba and Kimberly Knesting, "Zero Tolerance, Zero Evidence: An Analysis of School Disciplinary Practice," *New Directions for Student Leadership* 92 (2001): 17–43; and American Psychological Association Zero Tolerance Task Force, "Are Zero Tolerance Policies Effective in the Schools? An Evidentiary Review and Recommendations," *American Psychologist* 63, no. 9 (2008): 852–62.

137 **little evidence they do:** Derek W. Black, "Zero Tolerance Discipline Policies Won't Fix School Shootings," *Conversation*, March 15, 2018.

137 **Even students who are not suspended:** Brea L. Perry and Edward W. Morris, "Suspending Progress: Collateral Consequences of Exclusionary Punishment in Public Schools," *American Sociological Review* 79, no. 6 (2014): 1067–87. One might imagine that students' disruptive behavior causes *both* suspensions and declines in remaining students' grades, loss of trust in school officials, and so on. Perry and Morris,

though, controlled for disruptive behavior in their analyses. Their results demonstrate that even in schools with similar disruptive student behaviors, exclusionary discipline worsens school climate for nondisciplined kids.

137 **three times more likely:** Poetically, some of the best writing on this subject comes from Jason himself. For instance, Jason A. Okonofua et al., "A Vicious Cycle: A Social-Psychological Account of Extreme Racial Disparities in School Discipline," *Perspectives on Psychological Science* 11, no. 3 (2016): 381–98.

137 **Troublesome students are often troubled:** Albert Reijntjes et al., "Prospective Linkages Between Peer Victimization and Externalizing Problems in Children: A Meta-Analysis," *Aggressive Behavior* 37, no. 3 (2011): 215–22; and Kee Jeong Kim et al., "Reciprocal Influences Between Stressful Life Events and Adolescent Internalizing and Externalizing Problems," *Child Development* 74, no. 1 (2003): 127–43.

139 **When teachers believed the child was white:** Jason A. Okonofua and J. L. Eberhardt, "Two Strikes: Race and the Disciplining of Young Students," *Psychological Science* 26, no. 5 (2015): 617–24.

140 **simple exercise cut the racial gap:** Geoffrey L. Cohen et al., "Reducing the Racial Achievement Gap: A Social-Psychological Intervention," *Science* 313, no. 5791 (2006): 1307–10.

140 **a full "kindness curriculum":** Lisa Flook et al., "Promoting Prosocial Behavior and Self-Regulatory Skills in Preschool Children Through a Mindfulness-Based Kindness Curriculum," *Developmental Psychology* 51, no. 1 (2015): 44–51.

140 **A recent review of more than two hundred studies:** Joseph A. Durlak et al., "The Impact of Enhancing Students' Social and Emotional Learning: A Meta-Analysis of School-Based Universal Interventions," *Child Development* 82, no. 1 (2011): 405–32.

140 **work less well in older kids:** David S. Yeager, "Social and Emotional Learning Programs for Adolescents," *Future of Children* 27, no. 1 (2017): 73–94.

140 **Adolescents conform to each other:** Early adolescents (ten to four-teen years old) are especially compliant to peer norms. See Laurence Sternberg and Kathryn Monahan, "Age Differences in Resistance to Peer Influence," *Developmental Psychology* 43, no. 6 (2007): 1531–43.

141 **norms tend to win:** There are many other cases in which highlight-ing "bad" norms actually makes people more likely to misbehave. See, for instance, P. Wesley Schultz et al., "The Constructive, De-structive, and Reconstructive Power of Social Norms," *Psychological Science* 18, no. 5 (2007): 429–34.

141 **DARE does not appear to have reduced:** Chudley E. Werch and Deborah M. Owen, "Iatrogenic Effects of Alcohol and Drug Preven-tion Programs," *Journal of Studies on Alcohol* 63, no. 5 (2002): 581–90.

141 **Disciplinary problems plunged:** Elizabeth Levy Paluck et al., "Changing Climates of Conflict: A Social Network Experiment in 56 Schools," *Proceedings of the National Academy of Sciences* 113, no. 3 (2016): 566–71.

141 **their peers' empathy:** Erika Weisz et al., "A Social Norms Interven-tion Builds Empathic Motives and Prosociality in Adolescents" (in preparation).

142 **After teachers learned about empathic discipline:** Jason A. Oko-nofua et al., "Brief Intervention to Encourage Empathic Discipline Cuts Suspension Rates in Half Among Adolescents," *Proceedings of the National Academy of Sciences* 113, no. 19 (2016): 5221–26.

CHAPTER 7: THE DIGITAL DOUBLE EDGE

144 **Bilal recalls:** Wafaa Bilal, *Shoot an Iraqi: Art, Life, and Resistance Under the Gun* (San Francisco: City Lights Books, 2013).

145 **By 2017, that had ballooned:** Adam Alter, *Irresistible: The Rise of Ad-dictive Technology and the Business of Keeping Us Hooked* (New York: Penguin, 2017).

145 **This worries many people:** See, for instance, Sherry Turkle, *Alone Together: Why We Expect More from Technology and Less from Each*

Other (New York: Basic Books, 2017); and Jean M. Twenge, *iGen: Why Today's Super-Connected Kids Are Growing Up Less Rebellious, More Tolerant, Less Happy—and Completely Unprepared for Adulthood—and What That Means for the Rest of Us* (New York: Atria, 2017).

146 **Those who shared their experiences online:** Diana I. Tamir et al., "Media Usage Diminishes Memory for Experiences," *Journal of Experimental Social Psychology* 76, no. 1 (2018): 61–168; and Adrian F. Ward et al., "Brain Drain: The Mere Presence of One's Own Smartphone Reduces Available Cognitive Capacity," *Journal of the Association for Consumer Research* 2, no. 2 (2017): 140–54.

146 **Robert Vischer coined the term *Einfühlung*:** Robert Vischer, *Über das optische Formgefühl: Ein Beitrag zur Ästhetik* (Leipzig: Credner, 1873).

146 **In a cheery 2014 op-ed:** Mark Zuckerberg, "Mark Zuckerberg on a Future Where the Internet Is Available to All," *Wall Street Journal,* July 7, 2014.

147 **The more time we spend with people:** Linda Stinson and William Ickes, "Empathic Accuracy in the Interactions of Male Friends Versus Male Strangers," *Journal of Personality and Social Psychology* 62, no. 5 (1992): 787–97; and Meghan L. Meyer et al., "Empathy for the Social Suffering of Friends and Strangers Recruits Distinct Patterns of Brain Activation," *Social Cognitive Affective Neuroscience* 8, no. 4 (2012): 446–54.

147 **Has this reduced our ability to connect?** For a great summary of the effects of technology on empathy, see Adam Waytz and Kurt Gray, "Does Online Technology Make Us More or Less Sociable? A Preliminary Review and Call for Research," *Perspectives on Psychological Science* 13, no. 4 (2018): 473–91. For dehumanization over text, see Juliana Schroeder et al., "The Humanizing Voice: Speech Reveals, and Text Conceals, a More Thoughtful Mind in the Midst of Disagreement," *Psychological Science* 28, no. 12 (2017): 1745–62.

148 **more outrage when skimming the Internet:** Wilhelm Hofmann et al., "Morality in Everyday Life," *Science* 345, no. 6202 (2014): 1340–43.

148   **Trolls spend vast amounts of time:** In a fascinating study, psychologists found that Internet trolls are relatively *high* in cognitive empathy—understanding others' emotions—but unlikely to share their emotions. This allows trolls to apply savvy assessments of others when trying to generate content that will hurt them the most. See Natalie Sest and Evita March, "Constructing the Cyber-Troll: Psychopathy, Sadism, and Empathy," *Personality and Individual Differences* 119 (2017): 69–72.

149   **teenagers who are cyberbullied:** Mitch Van Geel et al., "Relationship Between Peer Victimization, Cyberbullying, and Suicide in Children and Adolescents: A Meta-Analysis," *JAMA Pediatrics* 168, no. 5 (2014): 435–42.

149   **tends to leave people more depressed:** Philippe Verduyn et al., "Passive Facebook Usage Undermines Affective Well-Being: Experimental and Longitudinal Evidence," *Journal of Experimental Psychology: General* 144, no. 2 (2015): 480–88.

149   **encourages us to broadcast fury:** For a powerful perspective on Internet outrage, see Molly J. Crockett, "Moral Outrage in the Digital Age," *Nature Human Behaviour* 1, no. 11 (2017): 769–71. For data on retweets of emotional and moral language, see William J. Brady et al., "Emotion Shapes the Diffusion of Moralized Content in Social Networks," *Proceedings of the National Academy of Sciences* 114, no. 28 (2017): 7313–18.

149   **Informational democracy disrupted national democracy:** Media theorist Zeynep Tufekci described the moment aptly on Twitter: "It's no longer age of information scarcity. Censorship works by info glut, distraction, confusion and stealing political focus and attention."

150   **"Dopamine makes your app addictive":** Jonathan Shieber, "Meet the Tech Company That Wants to Make You Even More Addicted to Your Phone," *TechCrunch*, September 8, 2017. Boundless Mind's materials no longer feature this motto, but it can be found on previous iterations of their website, for instance https://web.archive.org/web/20180108074145/https://usedopamine.com/.

151 **the number of homeless people has increased:** Jacqueline Lee, "Palo Alto Sees 26 Percent Rise in Homelessness," *Mercury News,* July 13, 2017.

151 **one announces over the bus's intercom:** See Elizabeth Lo, "Hotel 22," *New York Times,* January 28, 2015.

152 **activated by every group *except* the homeless:** Lasana T. Harris and Susan T. Fiske, "Dehumanizing the Lowest of the Low: Neuroimaging Responses to Extreme Out-Groups," *Psychological Science* 17, no. 10 (2006): 847–53.

152 **"VR is far more psychologically powerful":** Jeremy Bailenson, *Experience on Demand: What Virtual Reality Is, How It Works, and What It Can Do* (New York: W. W. Norton, 2018).

153 **these experiences decrease stereotyping:** For instance, Sun Joo (Grace) Ahn et al., "The Effect of Embodied Experiences on Self-Other Merging, Attitude, and Helping Behavior," *Media Psychology* 16, no. 1 (2013): 7–38; and Soo Youn Oh et al., "Virtually Old: Embodied Perspective Taking and the Reduction of Ageism Under Threat," *Computers in Human Behavior* 60 (2016): 398–410.

155 **technology also created longer-lasting empathy:** Fernanda Herrera et al., "Building Long-Term Empathy: A Large-Scale Comparison of Traditional and Virtual Reality Perspective-Taking" (2018), *PLoS One* 13, no. 10: e0204494.

156 **one feature typical of autism:** Autism typically affects cognitive empathy—understanding what others feel—while leaving other parts of empathy intact. For instance, people with autism exhibit brain mirroring, and vicariously take on others' emotions. For more, see Ian Dziobek et al., "Dissociation of Cognitive and Emotional Empathy in Adults with Asperger Syndrome Using the Multifaceted Empathy Test (MET)," *Journal of Autism and Developmental Disorders* 38, no. 3 (2008): 464–73; and Nouchine Hadjikhani et al., "Emotional Contagion for Pain Is Intact in Autism Spectrum Disorders," *Translational Psychiatry* 4, no. 1 (2014): e343.

156   **will never be able to improve:** Many people with autism do not *want* any more empathy than they have, and they thrive in ways people without it can't, for instance in lines of work that require great attention to detail. For much more on autism's history and contemporary advocacy, see Steve Silberman, *Neurotribes: The Legacy of Autism and the Future of Neurodiversity* (New York: Penguin, 2015).

157   **Critics of ABA:** Moreover, they argue that ABA focuses not on improving the experiences of people with autism, but rather on making them behave in ways that make *others* (such as their families, teachers, or coworkers) more comfortable. It does not help that Ivar Lovaas, ABA's inventor, also used behavioral techniques in cruel experiments, for instance, to "cure" boys of feminine characteristics.

157   **even "curing" some children:** See, for instance, Alyssa J. Orinstein et al., "Intervention for Optimal Outcome in Children and Adolescents with a History of Autism," *Journal of Developmental and Behavioral Pediatrics* 35, no. 4 (2014): 247–56.

157   **Mind Reading and similar programs:** Ofer Golan and Simon Baron-Cohen, "Systemizing Empathy: Teaching Adults with Asperger Syndrome or High-Functioning Autism to Recognize Complex Emotions Using Interactive Multimedia," *Developmental Psychopathology* 18, no. 2 (2006): 591–617.

158   **getting it right 75 to 90 percent of the time:** Soujanya Poria et al., "A Review of Affective Computing: From Unimodal Analysis to Multimodal Fusion," *Information Fusion* 37 (2017): 98–125.

158   **Affective computing is growing rapidly:** Its global market is projected to grow from $9.3 billion in 2015 to $42.5 billion in 2020; see Richard Yonck, "Welcome to the Emotion Economy, Where AI Responds to—and Predicts—Your Feelings," *Fast Company,* February 3, 2017.

158   **his cousin David:** David's name has been changed to protect his privacy.

160   **The Stanford team's preliminary results:** Jena Daniels et al., "5.13 Design and Efficacy of a Wearable Device for Social Affective

Learning in Children with Autism," *Journal of the American Academy of Child and Adolescent Psychiatry* 56, no. 10 (2017): S257.

163   **Mechanical Turk (MTurk for short):** MTurk is named after "the Turk," an eighteenth-century hoax. The Turk was a box about the size of a dining room table helmed by a mechanical statue that looked like a genie you might see at a Coney Island arcade. Supposedly, underneath it was an automaton that could play a decent game of chess—a harbinger of Deep Blue. In fact, a chess master hid inside the machine, making decisions for it. What appeared to be artificial intelligence was actually just the old-fashioned kind.

163   **crowdsourced editing:** Michael S. Bernstein et al., "Soylent: A Word Processor with a Crowd Inside," *Communications of the ACM* 58, no. 8 (2015): 85–94.

164   **turned to online communities:** Neil Stewart Coulson et al., "Social Support in Cyberspace: A Content Analysis of Communication Within a Huntington's Disease Online Support Group," *Patient Education and Counseling* 68, no. 2 (2007): 173–78; and Priya Nambisan, "Information Seeking and Social Support in Online Health Communities: Impact on Patients' Perceived Empathy," *Journal of the American Medical Informatics Association* 18, no. 3 (2011): 298–304.

166   **minutes of training sharpened helpers' empathy:** Robert R. Morris and Rosalind Picard, "Crowdsourcing Collective Emotional Intelligence," *arXiv preprint arXiv:1204.3481* (2012).

166   **Both groups experienced less depression:** Robert R. Morris et al., "Efficacy of a Web-Based, Crowdsourced Peer-to-Peer Cognitive Reappraisal Platform for Depression: Randomized Controlled Trial," *Journal of Medical Internet Research* 17, no. 3 (2015): e72.

166   **Generosity leaves givers fulfilled:** Elizabeth Dunn et al., "Spending Money on Others Promotes Happiness," *Science* 319, no. 5870 (2008): 1687–88; Cassie Mogilner et al., "Giving Time Gives You Time," *Psychological Science* 23, no. 10 (2012): 1233–38; and Peggy A. Thoits and Lyndi N. Hewitt, "Volunteer Work and Well-Being," *Journal of Health and Social Behavior* 42, no. 2 (2001): 115–31.

166 **especially true when givers experience empathy:** Sylvia A. Morelli et al., "Emotional and Instrumental Support Provision Interact to Predict Well-Being," *Emotion* 15, no. 4 (2015): 484–93; and Sylvia A. Morelli et al., "Neural Sensitivity to Personal and Vicarious Reward Differentially Relate to Prosociality and Well-Being," *Social Cognitive and Affective Neuroscience* 13, no. 8 (2018): 831–39.

167 **they became sharper empathizers:** Bruce P. Doré et al., "Helping Others Regulate Emotion Predicts Increased Regulation of One's Own Emotions and Decreased Symptoms of Depression," *Personality and Social Psychology Bulletin* 43, no. 5 (2017): 729–39.

167 **Trolls can spit venom:** When people send abusive messages on Koko, their text is used by Koko's algorithms to better prevent future misuse. Trolls inadvertently help the community keep them out!

EPILOGUE: THE FUTURE OF EMPATHY

168 **In a series of legendary speeches:** Terry Brighton, *Patton, Montgomery, Rommel: Masters of War* (New York: Crown, 2009).

169 **the iconic research of Walter Mischel:** Walter Mischel, *The Marshmallow Test: Mastering Self-Control* (New York: Little, Brown, 2014).

169 **Children who can't trust in promises:** Celeste Kidd et al., "Rational Snacking: Young Children's Decision-Making on the Marshmallow Task Is Moderated by Beliefs About Environmental Reliability," *Cognition* 126, no. 1 (2013): 109–14.

169 **so difficult to imagine them:** Richard L. Revesz and Matthew R. Shahabian, "Climate Change and Future Generations," *Southern California Law Review* 84 (2010): 1097–161.

170 **the earth should sustain life:** Nick Bostrom, "Existential Risk Prevention as a Global Priority," *Global Policy* 4, no. 1 (2013): 15–31.

170 **Singer suggests:** Peter Singer, "The Logic of Effective Altruism," *Boston Review,* July 6, 2015.

170 **On climate change, Bloom writes:** Bloom, *Against Empathy,* 112.

171   **gratitude protects against short-term thinking:** David DeSteno et al., "Gratitude: A Tool for Reducing Economic Impatience," *Psychological Science* 25, no. 6 (2014): 1262–67.

171   **more willing to sacrifice their own prosperity:** Kimberly A. Wade-Benzoni, "A Golden Rule over Time: Reciprocity in Intergenerational Allocation Decisions," *Academy of Management Journal* 45, no. 5 (2002): 1011–28.

172   **In the face of the universe:** Jennifer E. Stellar et al., "Self-Transcendent Emotions and Their Social Functions: Compassion, Gratitude, and Awe Bind Us to Others Through Prosociality," *Emotion Review* 9, no. 3 (2017): 200–207.

172   **people report feeling smaller, but more connected:** Michelle N. Shiota et al., "The Nature of Awe: Elicitors, Appraisals, and Effects on Self-Concept," *Cognition and Emotion* 21, no. 5 (2007): 944–63; and Paul K. Piff et al., "Awe, the Small Self, and Prosocial Behavior," *Journal of Personality and Social Psychology* 108, no. 6 (2015): 883–99.

172   **When people experience anxiety:** Lena Frischlich et al., "Dying the Right-Way? Interest in and Perceived Persuasiveness of Parochial Extremist Propaganda Increases After Mortality Salience," *Frontiers in Psychology* 6 (2015): 1222.

173   **people write their own eulogies:** When people consider their own legacies, they do tend to build kindness toward future generations. Kimberly A. Wade-Benzoni et al., "It's Only a Matter of Time: Death, Legacies, and Intergenerational Decisions," *Psychological Science* 23, no. 7 (2012): 704–9; and Lisa Zaval et al., "How Will I Be Remembered? Conserving the Environment for the Sake of One's Legacy," *Psychological Science* 26, no. 2 (2015): 231–36.

APPENDIX A: WHAT IS EMPATHY?

178   **Psychologists have (sometimes heatedly) debated:** Lauren Wispe, "The Distinction Between Sympathy and Empathy: To Call Forth a Concept, a Word Is Needed," *Journal of Personality and Social Psychology* 50, no. 2 (1986): 314–21; and Jamil Zaki, "Moving Beyond

Stereotypes of Empathy," *Trends in Cognitive Sciences* 21, no. 2 (2016): 59–60.

178   **It's an umbrella term:** That doesn't mean, by the way, that "empathy" is too vague a term to be useful. When you recall your first kiss, when you correctly access the year Pearl Harbor was attacked, when you accidentally autopilot yourself to work on a Saturday, you are experiencing three instances that can all be described as memory. Likewise, "empathy" encapsulates the ways other people's emotions affect us.

178   **These pieces:** Mark Davis, *Empathy: A Social Psychological Approach* (Boulder, Colo.: Westview, 1994); Simon Baron-Cohen and Sally Wheelwright, "The Empathy Quotient: An Investigation of Adults with Asperger Syndrome or High Functioning Autism, and Normal Sex Differences," *Journal of Autism and Developmental Disorders* 34, no. 2 (2004): 163–75; and Christian Keysers and Valeria Gazzola, "Integrating Simulation and Theory of Mind: From Self to Social Cognition," *Trends in Cognitive Sciences* 11, no. 5 (2007): 194–96.

179   **empathy's leading edge:** For a review of experience and mentalizing from evolutionary, developmental, cognitive, and neuroscientific perspectives, see Jamil Zaki and Kevin N. Ochsner, "Empathy," in *Handbook of Emotion*, ed. Lisa Feldman Barret et al., 4th ed. (New York: Guilford, 2016). *Einfühlung,* the term coined by German philosophers of art that was eventually translated to "empathy" by Edward Titchener, also closely resembles the modern definition of experience sharing.

179   **Smith, for instance, writes:** Adam Smith, *The Theory of Moral Sentiments* (Cambridge, UK: Cambridge University Press, 2002; first published in 1790).

180   **This cognitive piece of empathy:** See Alison Gopnik and Henry Wellman, "Why the Child's Theory of Mind Really Is a Theory," *Mind and Language* 7, nos. 1–2 (1992): 145–71; Chris L. Baker et al., "Rational Quantitative Attribution of Beliefs, Desires and Percepts in Human Mentalizing," *Nature Human Behavior* 1, no. 64 (2017); and Bill Ickes, *Everyday Mind Reading* (New York: Perseus Press, 2003).

180   **Concern has received less attention:** But see, for instance, Tania
        Singer and Olga M. Klimecki, "Empathy and Compassion," *Current
        Biology* 24, no. 18 (2014): R875–78.

180   **you must understand differences:** For examples of failures related
        to assuming similarity between self and others, see Nicholas Epley et
        al., "Perspective Taking as Egocentric Anchoring and Adjustment,"
        Journal of Personality and Social Psychology 87, no. 3 (2004): 327–39;
        Thomas Gilovich et al., "The Spotlight Effect in Social Judgment:
        An Egocentric Bias in Estimates of the Salience of One's Own Ac-
        tions and Appearance," *Journal of Personality and Social Psychology*
        78, no. 2 (2000): 211–22.

181   **activate different brain systems:** For review, see Jamil Zaki and
        Kevin N. Ochsner, "The Neuroscience of Empathy: Progress, Pitfalls,
        and Promise," *Nature Neuroscience* 15, no. 5 (2012): 675–80.

181   **Psychopaths have the opposite profile:** For instance, R. James Blair,
        "Responding to the Emotions of Others: Dissociating Forms of Em-
        pathy Through the Study of Typical and Psychiatric Populations,"
        Consciousness and Cognition 14, no. 4 (2005): 698–718.

181   **reliably increases our concern:** "Perspective-taking" exercises,
        which focus on mentalizing, are among the most well-tested ways
        to increase concern, at least in the short term. See C. Daniel Batson,
        Altruism in Humans (Oxford, UK: Oxford University Press, 2011).

181   **empathic processes promote kindness:** See, for instance, Zaki and
        Ochsner, "Empathy"; C. Daniel Batson and Laura Shaw, "Evidence
        for Altruism: Toward a Pluralism of Prosocial Motives," *Psychologi-
        cal Inquiry* 2, no. 2 (1991); and Michael Tomasello, *Why We Cooperate*
        (Cambridge, Mass.: MIT Press, 2009).

181   **"Russian Doll Model":** Frans de Waal, "Putting the Altruism Back
        into Altruism: The Evolution of Empathy," *Annual Review of Psychol-
        ogy* 5 (2008): 279–300.

181   **pointed not at any one individual:** This abstracted sort of concern,
        similar to what effective altruists cultivate, is also what René Bek-
        kers and his colleagues call the "principle of care"; see, for instance,

Mark O. Wilhelm and René Bekkers, "Helping Behavior, Dispositional Empathic Concern, and the Principle of Care," *Social Psychology Quarterly* 73, no. 1 (2010): 11–32.

APPENDIX B: EVALUATING THE EVIDENCE

183   **findings have proven less robust:** Open Science Collaboration, "Estimating the Reproducibility of Psychological Science," *Science* 349, no. 6251 (2015): aac4716; Andrew Chang and Phillip Li, "Is Economics Research Replicable? Sixty Published Papers from Thirteen Journals Say 'Usually Not,' " SSRN working paper, 2015; Leonard P. Freedman et al., "The Economics of Reproducibility in Preclinical Research," *PLoS Biology* 13, no. 6 (2015): e1002165; and Brian A. Nosek and Timothy Errington, "Reproducibility in Cancer Biology: Making Sense of Replications," *eLife* 6 (2017): e23383.

IMAGE CREDITS

22   Jamil Zaki.

23   Jamil Zaki.

33   Ron Haviv/VII/Redux. Used by permission.

33   Ed Kashi Photography LLC. Used by permission.

35   Kurt Lewin, "Group Decision and Social Change," in *Readings in Social Psychology*, ed. Guy Swanson et al. (New York: Henry Holt, 1952), 459–73.

57   Nour Kteily et al., "The Ascent of Man: Theoretical and Empirical Evidence for Blatant Dehumanization," *Journal of Personality and Social Psychology* 109, no. 5 (2015): 901.

178   Jamil Zaki.

INDEX

Data (char.), 10, 11, 12
data harvesting, social media and, 149–50
Davos World Economic Forum, 153
dehumanization, 116
    contact as antidote to, 64
    and decrease in empathy, 25, 56–57, 68
delayed gratification, 169, 171
#depression, 165
d'Er Roh, France, 172
Descent of Man, The (Darwin), 4
DeSteno, David, 26–27, 171
detached concern, 102, 112
Detroit, Mich., 1943 race riot in, 60–61
de Waal, Frans, Russian Doll Model of, 181
diet, human evolution and, 15
diseases, online communities and, 164, 195
divorce, of author's parents, 1–3, 16
Domestic Tension (Bilal gallery installation),
    144–45
Dopamine Labs (Boundless Mind), 150
drones, pilots of, 144
Dweck, Carol, 27–28, 49
    empathy experiments of, 29–31
    mindset research of, 28–29, 73

Eastern philosophies, emotions in, 13
effective altruism (EA), 169–70
Einfühlung (feeling into), 146
Ekman, Eve, 111, 113, 116
Ekman, Paul, 111, 113
election, U.S., of 2016, Russian
    misinformation campaign in, 149
Ella (YPT actor), 77–78, 81
Emotient, 158
emotional detachment, 37–38, 52, 55, 68
    cultivation of, 34
emotional empathy, 4, 178
emotional granularity, 112–13
emotional intelligence (EI), 12, 13
emotions:
    contagiousness of, 38, 179
    cultivation of specific, 38, 189
    mirroring of, 14
    as products of thought, 13, 37
    psychological tuning and, 38
    viewed as instinctive reflexes, 12–13, 36
empathic concern, 4, 113–14, 116, 178,
    180–82
    metta meditation and, 115
empathic distress, 113–14, 115
    negative effects of, 114
empathic listening, 161
empathy:
    in animals, 5, 6

brain regions activated by, 181
caregivers and, 94–95, 102–5, 112, 113–18,
    190
as caring, 178, 180–82
communities and, 7, 41
compassion fatigue and, 95–97
as contagious, 13
definition of, 4, 178–82
effect of experience on, 23–27
emotional, 4, 178
as eroded by inflicting harm or
    unhappiness, 24–25, 188
excessive, 96, 103, 112, 113, 116–18
expanding reach of, 6–7, 43–44, 63, 146,
    169–70, 181
genetic inheritance and, 22, 188
hate groups and, 71
health costs of, 96–97
human evolution and, 5–6, 41, 187
as increased by contact with outsiders,
    61–63, 191
as increased by suffering, 24, 25–27, 188
toward individuals vs. groups, 8–9, 43,
    63, 187
isolation and, 41
kindness and, 4, 5, 14, 187
as lacking toward homeless people, 152
as mental untethering, 75–76, 81
mirroring and, 14, 188
New Dawn and increase in, 85–87
in parents, 96
personal benefits of, 40–41
police officers and, 130, 131, 132
psychological tests for, 11–12
Russian Doll Model of, 181
as sharing, 178–79
social norms and, 120, 121, 122, 193
storytelling and, 76
as survival skill, 5, 173
as thinking, 178, 180–82
tribalism and, 9, 44–45
tuning of, 116–18
twin studies and, 22–23
as umbrella term, 178
viewed as fixed trait, 10–12, 14, 22, 156
viewed as reflex, 12–13, 14–15, 34, 36–37
in women vs. men, 44
empathy, choice and:
    AIDS epidemic and, 43–44
    avoidance and, 39–41, 189–90
    and feelings of being overwhelmed, 40, 41
    incentives and, 44
    Lewin's force theory applied to, 41, 43
    and moral self-image, 39, 189

empathy, choice and (*cont.*)
  negative impact of conflict on, 63
  nudges and, 42–45
  in psychopaths, 45–46
  relationship building and, 38–39, 189
  usefulness and, 38–39
empathy, cognitive, 4, 178, 180–82
  acting and, 79–81
  of children, 79–80
empathy, decrease in, 182
  competition and, 44–45, 191
  conflict and, 58, 63, 66, 182
  isolation and, 7–8
  in modern world, 7–10, 51
  moral disengagement and, 25
  online technology and, 147, 149, 150,
    195–96
  toward outsiders, 7, 37, 56–57, 68
  in political discourse, 9, 55–56
  racism and, 56–57
empathy, as learnable trait, 15–16, 170, 182
  adolescents and, 141
  effectiveness of believing in, 27, 30–32,
    49–51
  "fast-twitch" vs. "slow-twitch" changes
    in, 46–47, 49–51
  fiction reading and, 81–82, 192
  Koko and, 166–67
  long-term thinking and, 173, 196
  meditation and, 47–49, 190, 193
  mindset and, 29–32, 143, 188–89
  nudges and, 119
  police training and, 130, 131, 132, 193–94
  teachers and, 142–43, 194
  theater and, 79–81, 191–92
  VR technology and, 153–54, 195
empathy bias, 134–35, 194
Enos, Ryan, 62
Epictetus, emotions as viewed by, 13, 37
eugenics, 11
European Union, 73
ex-convicts, recidivism among, 87, 91, 192
Expanding Circle, The (Singer), 6, 181
experience sharing, 178, 179–80

Facebook, 149–50, 152, 164, 167
Ferguson, Mo., 132
fiction reading:
  empathy and, 81–82, 192
  and envisioning of future selves, 89–90
  see also Changing Lives Through
    Literature
fixists:
  everyday, 28–32, 46

psychological, 18–22, 46
Flynn, James, 21
formers (former hate group members),
    68–72
Francisco (premature baby), 99–100
  life support removed from, 105, 107, 117
Frankl, Victor, 27
future selves, envisioning of, 72–73, 89–90

gacaca courts, Rwandan genocide and, 84,
    86
Galileo Galilei, 17
Galton, Francis, 11, 20
gay rights, 55, 121
Gekko, Gordon (char.), 120
generosity, genetic inheritance and, 23
genes, survival of, 4–5
geology, fixists vs. mobilists in, 18
Goldstein, Thalia, 78–79, 92–93
  acting career of, 79
  cognitive empathy research of, 79–81
Good Samaritan, parable of, 40
Google, 122
Google Glass, 158–59
  autism and, 159–61
  commercial failure of, 155–56, 160
gratitude, long-term thinking and, 171, 196
"Greasy Lake" (Boyle), 88–89
Gross, James, 37–38
Grossman, Dave, 125–26, 131
Gun-Free Schools Act (1994), 136–37

Haber, Nick, 158–59
Habyarimana, Juvénal, 83
Haiti, 2010 earthquake in, 9
Hamlet (Shakespeare), 37
harm, infliction of, empathy as eroded by,
    24–25
Harvey, Hurricane, 26
Harwood Manufacturing, 41–42
hate groups, 87, 120
  child abuse victims in, 53, 69
  defense mechanisms of, 70–71
  empathy and, 71
  and estrangement from future selves,
    72–73
  former members of, 68–73
hatred, 116
  as defense mechanism, 60, 69
  mutation of, 55
  of outsiders, 55
  psychological roots of, 53
  see also prejudice; *specific prejudices*
Haviv, Ron, 33–35, 37

Hemingway, Ernest, 90–91
Herrera, Fernanda, VR experiment of, 154–55
Heyer, Heather, 68
HITs (Human Intelligence Tasks), 163
  Morris's use of, 164–65
Holmes, Stephanie, 77–78
homeless people:
  increased empathy for, 154–55
  lack of empathy for, 152
  in Santa Clara County, 151–52
homophobia, 52
Houston, Tex., 26
human evolution, empathy and, 5–6, 41, 187
humanity, existential threats to, 169–70
human nature:
  as combined result of inheritance and experience, 21
  storytelling as essential trait of, 76, 93
  viewed as fixed, 18–22
Hungary, 64
Hutu Power movement, 84
Hutus, 83, 84, 86

IDEO, 122
imagination, as mental untethering, 74–75
Instagram, 165
intelligence:
  effect of environment on, 21, 22, 188
  genetic inheritance and, 22
  as malleable, 29, 188
intensive care nurseries (ICNs), 117
  emotional investment of staff in, 102–3
  staff health problems and burnout in, 103–4
  staff reluctance to share experience of, 107
  see also Benioff Children's Hospital
Internet, see online technology
interracial marriage, general acceptance of, 55
isolation:
  effect of urbanization and shrinking household size on, 7–8
  empathy and, 41
Israelis, Palestinians and, 38, 63, 64, 67, 71–72, 73

Jim Crow South, 148
Johns Hopkins Hospital:
  Josie King Patient Safety Program at, 108
  RISE network at, 109–10
Joinson, Carla, 96
Josie King Patient Safety Program, 108

Journal of Geology, 18
Jungle, The (Sinclair), 82
Justice Statistics Bureau, 87

Kahneman, Daniel, 163
Kane, Bob, 87–88, 90, 91, 92
Kant, Immanuel, empathy as viewed by, 37
karuna, 180
Kashi, Ed, 33, 34–35, 37
Katrina, Hurricane, 26
Kelton, Fraser, 167
Keysers, Christian, 46
Kik, 167
kindness:
  as cost/benefit ratio vs. win-win situation, 166, 195
  empathy and, 4, 5, 14, 187
  fighting for, 16, 168–69
  mirroring and, 14
  social norms and, 120
  and survival of genes, 5
  as survival skill, 5, 173
  of trauma survivors, 26
King, Angela, 70, 71–72
King, Josie, 108, 109
King, Sorrel, 108
King, Tony, 108
King County Sheriff's Office, 122, 127
Knezovich, Ozzie, 132–33
Koko (social media bot):
  author's experience with, 161–62
  empathy increased by, 166–67
  Morris's creation of, 166
Kosovo War, ethnic cleansing in, 33
Kouddous, Kareem, 167
Kteily, Nour, 56–57, 69–72
Kurdi, Alan, 8–9

Lewin, Kurt:
  action research of, 41–42
  force theory of, 35–36, 38, 41
Liebowitz, Melissa, 98, 99, 100, 101–2, 103–4, 105–6, 107, 117–18
Life After Hate (nonprofit organization), 69, 87
  Northwestern University conference of, 69–72
Life After Hate (online journal), 68
Lim, Daniel, 26–27
Lincoln, Abraham, 82
Lipps, Theodor, 13
literature, viewed as superfluous, 87, 92
Liverpool (soccer team), 45
Living Library School Project, 64

thinking:
empathy as, 178, 180–82
about mortality, 172–73
short-term vs. long-term, 169–70, 171–73,
196
time, untethering from, 74–75
To Kill a Mockingbird (Lee), 155
trauma survivors, increased empathy of, 26
tribalism, 63
British soccer fans and, 45
empathy and, 9, 44–45
social media and, 149
Troi, Deanna (char.), 10, 11, 12
trolls, 9, 148, 149, 165, 167
Trump, Donald, 57
Trump administration, 137
Tutsis, 83, 84, 86
twin studies, empathy and, 22–23
Twitter, 149–50, 167
Tyler, Tom, 128

Uncle Tom's Cabin (Stowe), 82
United Kingdom, immigrants in, 62
untethering, mental, 74–75
activation of brain regions in, 75
empathy as, 75–76, 81
online technology and, 146
urbanization, isolation and, 7

Virtual Human Interaction Lab, 150–51
virtual reality (VR) technology:
empathy and, 153–54
learning facilitated by, 152–53
stereotyping and discrimination reduced
by, 153, 195
Vischer, Robert, 146
Vonnegut, Kurt, 74
Voss, Catalin, 158–59, 160
Voyager 1, 171

Wall, Dennis, 159
Wallach, Ari, 169, 170–71, 173
Wall Street (film), 120
Wall Street Journal, "Blue Feed, Red Feed"
project of, 147–48
WarGames (film), 3
Washington State Criminal Justice
Training Commission (CJTC), 130
classroom instruction at, 129

LEED training and, 128
mental illness and, 131
Mock City simulations at, 129–30
race-based training at, 131
Rahr appointed director of, 127
Rahr's redesigned curriculum for, 128–
30, 132
warrior mentality rejected by, 128
water shortages, 169
Waxler, Bob, 87–92
Wegener, Alfred, 17–18, 21
Weinstein, Harvey, 121
Weiss, George:
New Dawn created by, 83, 84–85
and success of *New Dawn,* 86
Weisz, Erika, 49–50, 141
VR experiment of, 154–55
White Aryan Resistance (WAR), 52–53, 68
white supremacy, white supremacists, 52,
53–55, 58, 59
as increasingly emboldened, 68
Winters, Joe, 129–30
Wisconsin, University of, Center for Healthy
Minds of, 140
Wu, Albert, 109–10

Yemen, humanitarian crisis in, 9
Young Performers Theatre (YPT), 76–77

Zaki, Alma, 94–95, 97, 117
Zaki, Jamil (author):
anxiety over book of, 161
empathic norms researched by, 121
empathy experiments of, 29–31
Koko and, 161–62
at Northwestern University conference,
69–72
and parents' divorce, 1–3
VR experiment of, 154–55
Zambrano-Montes, Antonio, police killing
of, 133
zero tolerance policies:
adversarial culture promoted by, 138
lack of evidence for efficiency of, 137
racial stereotypes and, 137, 138
teachers and, 138
Zuckerberg, Mark, 146, 167
Zuckerberg San Francisco General Hospital,
111

ABOUT THE AUTHOR

JAMIL ZAKI is a professor of psychology at Stanford University and the director of the Stanford Social Neuroscience Lab. Using tools from psychology and neuroscience, he and his colleagues examine how empathy works and how people can learn to empathize more effectively. His writing on these topics has appeared in the *New York Times*, the *Washington Post*, the *New Yorker*, and the *Atlantic*. He lives in San Francisco with his wife and their two daughters.